PRACTICAL GEOMETRIC PATTERN DESIGN

DECAGONAL PATTERNS IN PERSIAN TRADITIONAL ART

Fragment of a wall mosaic in the Chahar Bagh Theological School known also as School of the Four Gardens, Isfahan, Iran.

MIROSŁAW MAJEWSKI

PRACTICAL GEOMETRIC PATTERN DESIGN

DECAGONAL PATTERNS IN PERSIAN TRADITIONAL ART

ABOUT THE AUTHOR

Dr Mirosław Majewski

Former Professor of New York Institute of Technology. Currently NYIT Professor Emeritus. Author of about 50 papers and books on computer graphics, applications of computers in education, mathematics and computer science education, geometry in art and architecture. Some of his recent papers and books are related to the geometry in Central Asian traditional art and history of medieval mathematics.

E-mail: mirek.majewski@yahoo.com
Internet: symmetrica.wordpress.com and majewski.wordpress.com

© Mirosław Majewski

ISBN: 9798730179325

Imprint: Independently published

Second edition, 2021

Photo on the cover: a Tan Sahid style design

Layout, all drawings and all photographs in this book were created by the author.

FOREWORD

I have been researching geometric art and organizing various workshops and courses related to geometric patterns in Islamic, Chinese, and European arts for years. Sometimes my students were asking me about a textbook that they could use in my classes. I must admit I never had enough time to write such a book. Finally, in November 2018, I found a moment when my everyday duties were giving me a few weeks off. This was the right time to sit down and start typing.

In this book, there is not much talking. My explanations are very brief and just to the point. Drawings and sometimes photos are a significant part of this book. Simply I do not want to waste much time and space to tell stories or show lovely photographs. You can find them in numerous books and Internet resources. Geometric drawings are international. We do not need much explanation to read a picture, and in this book, images are enough to explain the concepts described here.

The whole book is devoted to the decagonal geometric patterns in Persian traditional arts and architecture. What means the word 'decagonal', I will explain it slightly later. Why decagonal? This is another question worth asking. I often visit Turkey, and Istanbul is my favorite place. In most of the mosques in Istanbul and the Topkapi Palace, we see mostly decagonal patterns in various styles. The same is in many places in Iran. Mosques and madrasah in Iran often contain decagonal patterns.

There is also another important reason for concentrating on decagonal patterns is geometry. All of them share the same or very similar features – local symmetries of a regular decagon and regular pentagon. The golden section is a common feature of most of decagonal designs. For a student, it is much easier to concentrate on one type of geometry and follow it through many geometric constructions and designs.

The first edition of this book was published by Istanbul Design Center in 2019 under the title "Practical Geometric Pattern Design, decagonal patterns in Islamic art (part 1). The draft for the second edition was ready at the beginning of 2020. Meanwhile, a few things changed.

I usually teach my courses with Geometer's Sketchpad as a tool for demonstrating material. At the same time, my students enjoy drawing patterns by hand with compasses, rulers, pencils, and a lot of coloring tools. In July 2019, it was announced that McGraw Hill no longer provides Geometer's Sketchpad software or Geometer's Sketchpad licenses. Thus GSP should be considered as free software and used with any available license key. This means anybody can use it provided that he knows the location to download the software and has a license for it. Both – the software and the license key – can be easily found on the internet. You need to search for them via Google or any other searching software.

Another important thing that came out almost the same was the pandemic and necessity to move from face-to-face teaching courses to the online Zoom and Jitsi supported lectures. In such an environment, Geometer's Sketchpad proved to be an ideal tool for such teaching mode. In the courses that we had at IDC in fall 2020 and spring 2021, we used GSP for my presentations and students' works. Our students use Geometer's Sketchpad for learning pattern design and ruler and compasses to make their artworks.

The impact of these two factors forced me to revise my book for readers using compasses and ruler drawing techniques and software like Geometer's Sketchpad. In this book, there is a very brief section introducing Geometer's Sketchpad for constructing patterns. A more elaborated document can be found on my website:

https://symmetrica.wordpress.com

This is where readers of this book can find some additional examples and tutorials. This is also the place where you will find links to Geometer's Sketchpad download sites and the license key for the Windows version of GSP. Sketchpad for Mac no longer needs any license key.

I am grateful to all those who have helped me in many ways during my research and writing, particularly Mehmet Husrev Seref, to support this book's idea and all my Turkish adventures. I am very grateful to my students and staff of the Istanbul Design Center for all attention and support I got from them.

I wanted to express my personal gratitude to my Iranian friends, Masoud Mousavizade, Farnoosh Daneshpanah, and Zohor Mousvizade, for the effort and extra time you have contributed to my Iranian trips and my research on Persian architecture.

I am particularly grateful to Anthony Lee. His notes and email comments inspired me frequently in my research on geometric pattern design.

I am also indebted to the late G. I. Gaganov, who put my research on track of Central Asian geometric pattern design.

TABLE OF CONTENTS

Foreword 1
Table of contents 3

Chapter 1. Gereh and its rules 5
   Scrolls of medieval architects... 5
   Gereh rules 10
   Some necessary explanations... 11

Chapter 2. Selected geometric constructions..13
   Using Geometer's Sketchpad in pattern design 19

Chapter 3. Styles of decagonal patterns..21
   Project 3.1 – Our first contour and tessellation 22
   Project 3.2 – The Nodir Devon Madrasah style.. 23
   Project 3.3 – The Kukeldash Madrasah style 24
   Project 3.4 – The Persian style 26
   Project 3.5 – Tan Sahid Mosque style 28
   Project 3.6 – Kites and roses 30
   Project 3.7 – Kites and roses 2.. 32
   Project 3.8 – A decagonal mashrabiya 34
   Project 3.9 – A medallion from Şehzade Mosque 34
   Project 3.10 – The Sikandra style 36
   Project 3.11 – Sikandra style with Mashhad twist 37
   Project 3.12 – Decagonal pattern Samarkand style...................................... 38
   Project 3.13 – Multiple tessellations for one pattern 40
   Chapter Summary.. 43

Chapter 4. Patterns on decagonal grids 45
   Project 4.1 – Interlace patterns... 45
   Project 4.2 – Interlace patterns piece by piece.. 47
   Project 4.3 – Banding Mashhad style 43
   Project 4.4 – Simple decagonal grid pattern Tony Lee style......................... 44
   Project 4.5 – The Yazd decagonal grid... 53
   Project 4.6 – The Yazd style swirl with stars 54
   Project 4.7 – The Yazd style swirl with stars and loops................................. 56
   Project 4.8 – The Yazd style with some extensions 57
   Project 4.9 – A monster Yazd style pattern 60

Chapter 5. Proportions, contours, and tessellations.. 61
   Decagonal contours 62
   Selected basic decagonal tessellations 65
      Group A. The most frequently used tessellations................................... 65
      Group B. Tessellations with long contours.. 70
      Group C. Tessellations using combined contours.................................. 71
      Group D. Tessellations with long kite... 73
      Group E. The tangent and overlapping decagons method 73
      Group F. Tessellations using a single tile 78
      Group G. Tessellation with narrow rhombi 79
   Properties of polygons used in decagonal tessellations............................... 80

Chapter 6. Kukeldash Madrasah and Persian styles 81
Elements of the Kukeldash Madrasah style 81
Project 6.1 – Pattern from Cem Sultan Mausoleum 84
Project 6.2 – Pattern from Ankara Ethnographic Museum 86
Project 6.3 – Pattern from stone in Amasya 86
Project 6.4 – Pattern Beyazid Mosque in Amasya 88
Project 6.5 – Pattern from the Great Mosque in Damascus 89
Elements of the Persian style 91
Project 6.6 – Pattern from Beyazid II Mosque in Istanbul 92
Project 6.7 – Pattern from Muradiye Mosque in Bursa 93
Project 6.8 – Tessellation with four hexagons 94
Project 6.9 – Doors from the David Collection 95
Project 6.10 – An experimental Persian design 98

Bibliography 101

1. Gereh and its rules

As we know, there are numerous books and papers on geometric patterns in Islamic arts. There are also numerous methods of designing such patterns. In many Muslim countries, there are local methods, usually based on specific designing techniques relevant to the material used in rendering the patterns, e.g., the hasba method for woodworking, zellige for ceramic tiles, etc. These methods often look into the geometry of specific shapes used in these patterns. There are also western methods utilizing some kind of geometric approach. In this book, we will concentrate on the gereh method that has its origins in Central Asia. The word *gereh* (knot) is an Iranian term. They say *gereh sazi* (tying knots) and *gereh chini* (the art of making knots). As we shall see later, the word 'knot' describes precisely how these patterns are created.

Scrolls of medieval architects

Before we dive into geometry and construction details, it would be wise to look at some resources where we can find reliable information about geometric patterns. First, and the most important – there are no books or written documents from the Middle Ages describing geometric pattern design techniques. However, we are not entirely without any information from this period. We have many patterns in mosques and other places. We also have a few scrolls of architects from the middle ages and later times. Each such scroll is a valuable source of information if we know how to read it.

The most important, at the moment, is the Topkapi scroll. It is a long roll of glued parchment pieces with 114 images. Its length is 29.5 meters and 33 cm wide. We suppose that the scroll was created during the Safavid dynasty in Iran by an unknown architect or designer. Currently, the scroll is deposited in the Topkapi Museum, which takes its name from it, and it is not on display. However, several images from this scroll are scattered on the Internet.

Other known scrolls are stored in the Victoria and Albert Museum. This is a collection of architectural sketches forming a *"portfolio of 238 designs on paper, once owned by a working architect in Qajar Tehran, in nineteenth-century Iran. There are two complete paper scrolls, and 236 smaller designs, most of which were cut down from other scrolls. They are a rare survival. The drawings vary in style and content, showing a range of designs proposed for tilework, stucco, and woodwork, as well as architectural ground plans and elevations. Some reflect Iranian traditions of long-standing, while others show decorative fashions imported from Europe. They are probably the work of several different individuals.*

> **MEMO**
> Scrolls of architects are collections of drawings representing patterns, plans of architectural components or even whole buildings. Usually they were kept in a family for many generations.

The drawings were acquired for the Museum in 1875 by Caspar Purdon Clarke, an architect who later became Director of the V&A. In 1874-75, Purdon Clarke was in Tehran, renovating the British embassy buildings. During the project, this drawing series was presented to Purdon Clarke by the local master-builders he was working with. He reported later that this was not a sale but an exchange, in acknowledgment of his teaching some European building-techniques to his Tehran colleagues. The two master-builders, Ostad Khodadad, and Ostad Akbar explained that the portfolio had belonged to the late Mirza Akbar, a court architect active in Tehran earlier in the century." (This information was taken from the V&A website).

Finally, there are scrolls of architects in Uzbekistan museums. How many? In his book (1961), Rempel remarks on a few of them without referring to their inventory numbers. In 1936 M. S. Andreev went to Bukhara, where he collected 8 scrolls with geometric patterns and muqarnas plans. These scrolls were deposited in the Bukhara Museum (which one?) and were never described in detail. In his short paper (Andreev, 1958), he wrote a very brief description of these scrolls and enclosed the inventory numbers of some of them. But this is practically all that we know about them. In 2018 I was able to examine photos of one of these scrolls. But the quality of the designs was relatively poor, and the designs were very basic. Perhaps one day we will be able to examine all of them? Meanwhile, let us concentrate on what is available online from the Mirza Akbar scrolls and the Topkapi scroll.

> **MEMO**
> A template for a pattern is a reasonably small part of it, usually rectangular or sometimes triangular, that can be used to recreate the whole pattern.

Each of these scrolls is a collection of rectangular drawings. In very few cases, the pattern is triangular. Each drawing is a separate template for a pattern or a plan of muqarnas. As we will see later, each template has a particular proportion, and it usually presents a quarter of a pattern. Here is one example reconstructed from the Topkapi scroll.

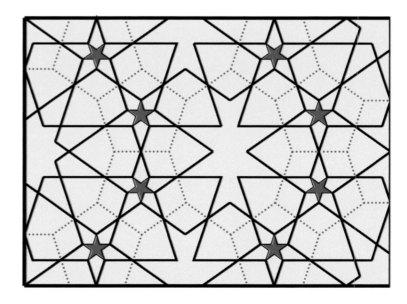

Fig. 1.1. Fragment of the Topkapi scroll reconstructed by the author. Here we see all parts of the design, including some extra decorations. The small stars were added to make the whole design more attractive. However, they do not follow the general rules of the entire pattern. Thus while reconstructing this pattern, we can ignore them or add them at the end of the designing process.

A specific feature of this drawing is that we deal here with the contour (a rectangle), the pattern (black lines), and a polygonal network marked here by red dotted lines. Thus we have three elements – a contour, a network of dotted lines usually called a tessellation of the pattern, and the pattern. These three elements we will call a gereh. When we see a gereh rendered in wood or ceramic, we see only the pattern. The dotted or red lines are usually omitted. In many templates from the Topkapi scroll, the polygonal network is typically red, and its lines are dotted or continuous.

> **MEMO**
> Gereh = contour + tessellation + pattern
> The tessellation or network is the geometry driving the design of the pattern.

Each of these three elements will be explained in detail in a moment. Before doing so, let us think about how we can create a larger pattern using such a template. For this purpose, we should forget for a moment the dotted lines. This way, we will have only the contour and the pattern inside this contour. We should use multiple copies of the template. How?

The following picture shows what will happen if we simply take two copies of the template and put them next to each other. The red circles show the places where the pattern is not aligned.

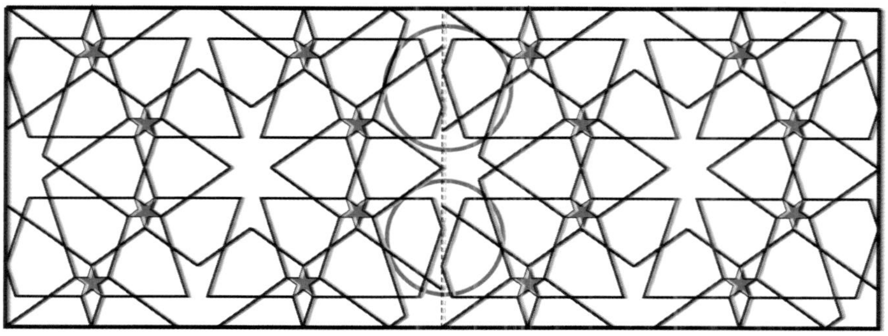

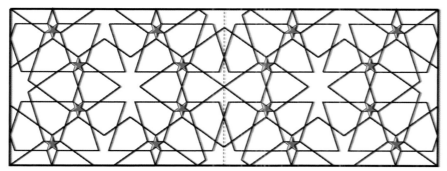

Fig. 1.2. This is the wrong way of creating a pattern from a template shown in fig. 1.1. Here we clearly see that the middle part of the design has lost its continuity. This certainly offends our sense of aesthetics. A viewer would expect that the lines of the pattern should flow smoothly without any breaks.

Therefore we should use another approach. Draw the first instance of the template, and then next to it, draw its mirror reflection about one of the sides – here left or right.

Fig. 1.3. The template and its mirror reflection make a perfect pattern without any unwanted breaks of lines.

In figure 1.3, we see how the template from figure 1.1, and its mirror reflection about the right side, fit together. If we remove the dashed line separating the left and right sides, we will not see where these two copies of the template are joined.

One can quickly notice that in this way, we can produce a few different patterns using one template only. It is enough to take a few copies of the template and its mirror reflections to create something that fulfills our needs.

> **MEMO**
> One template can be used to create a few patterns that may look very different.

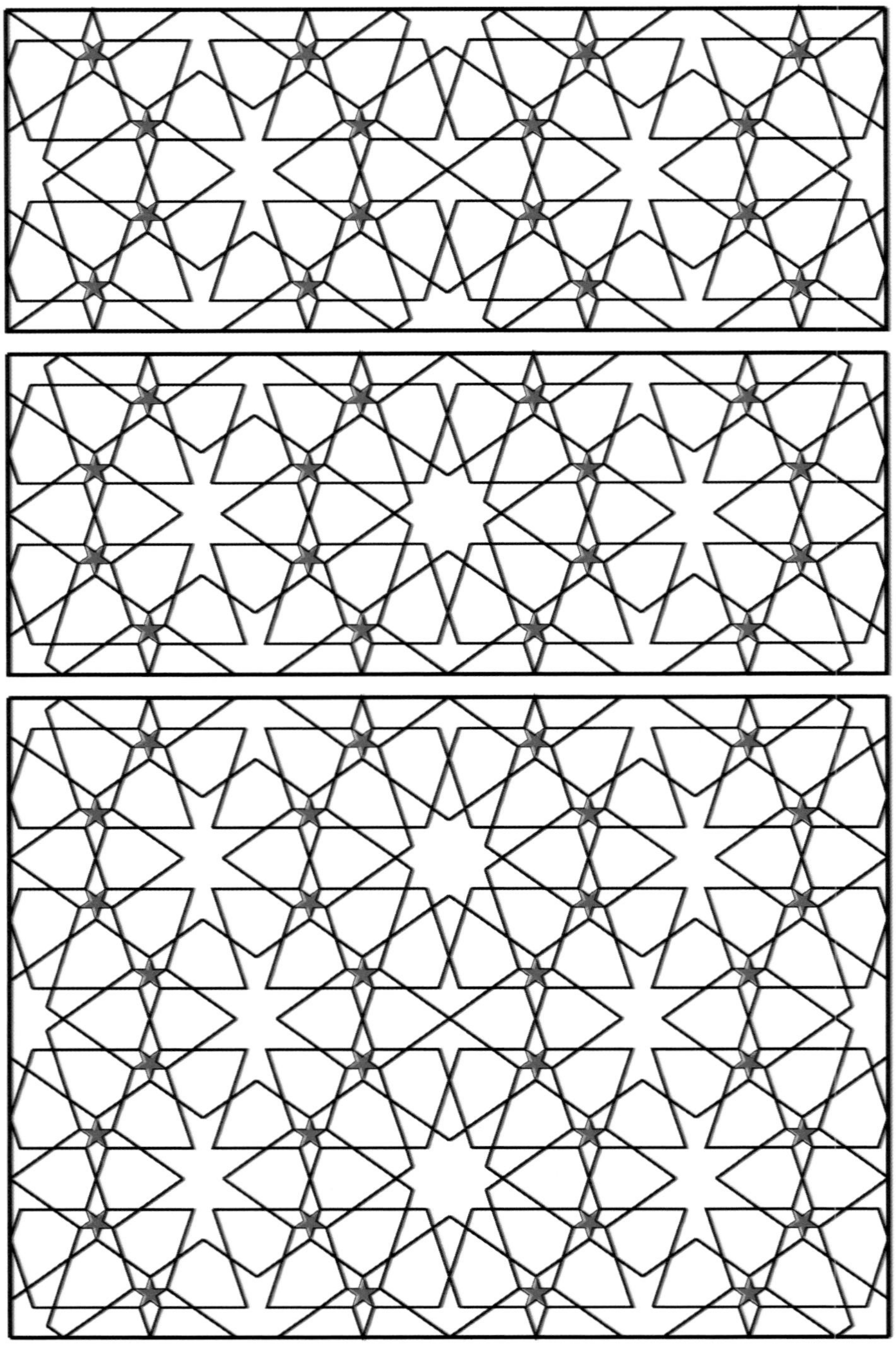

Fig. 1.4 Here, we show three ways of creating patterns from the same template. In the top one, our focus is placed on a star-like shape that is a very unusual element of a decagonal design. We will talk about it later.

The middle image is a typical solution that we often see in real applications. Most of the kundekari doors in Turkey, as well as in other places, have a similar look. Here the central star is the main focal point. This is often the desired artifact.

The large pattern in the bottom image was created using eight copies of the same template or its mirror reflections. An image like this has many focal points.

From all these images, we can conclude that the network of dotted lines on the template only makes a geometric framework for the pattern, and it should be removed after finishing the complete design.

Therefore, what is the purpose of the red network? We will refer to it as a tessellation of the pattern. In Russian literature, it is called 'a network of a pattern'. The tessellation is the geometry behind the pattern, and the pattern itself is only a decoration of this geometry. As one can easily conclude here – each tessellation may have many different decorations.

While looking at images from the Topkapi scroll, one can make the impression that the creation of a geometric pattern is like assembling a puzzle from decorated tiles. Many people do this without bothering how the patterns were constructed in the past. The enclosed photo (figure 1.5) shows a scroll published by Rempel (see Rempel, 1961, page 402). The picture shows a network of white lines that are not seen on Topkapi scroll or even on Mirza Akbar scrolls. Why? It is easy to guess that a photograph of a paper taken in intense front light will not show such elements. We need a light from the side that will allow us to take a better photo. What are the white lines? Here is the explanation from the V&A website "*Each design was constructed using a compass and dry-point ruler: the initial construction lines remain scored into the paper, as does the hole in the centre from the designer's compass-point.*"

> **MEMO**
> Each design was constructed using a compass and dry-point ruler: the initial construction lines remain scored into the paper, as does the hole in the center from the designer's compass-point.

The photo from Rempel's book also shows the network of polygons. In the Topkapi scroll, the edges of polygons were marked with a red pen. Here they are left in the form of impressed lines. Only the white trace shows where we should look for them.

One can easily notice that the white lines show how the contour and tessellation were constructed. We will come back later to this drawing, and we will reconstruct the whole pattern using the white traces.

Fig. 1.5. The template of a pattern with visible construction lines (photo from Rempel, 1961, page 402). This photo shows how the design was created. The contour and locations of particular stars were constructed by divisions of particular angles into 5 or 10 equal parts. The division lines also show how the polygons of the tessellation were created. Note – the regular decagons here are drawn with all their symmetry lines. That is quite useful while designing stars and rosettes of the pattern.

In Rempel's book, there are four photos taken from the architects' scrolls. Each of them shows the same features – white construction lines, tessellation, and the pattern. According to Rempel, each of these drawings comes from an XVI century scroll.

At the moment, we have enough examples and sufficient information to create a clear and precise theory of gereh. Our approach will be developed based on eight rules formulated like any other axiomatic system in mathematics.

THE GEREH RULES

Let us start by analyzing the same template that we saw in figure 1.1 but with some extra markings.

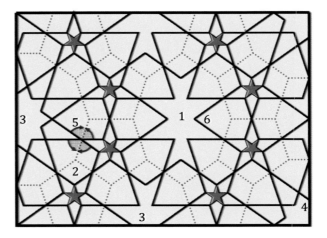

Fig. 1.6. Fragment of the Topkapi scroll reconstructed by the author. The numbers added here refer to the specific rules listed below.

Below we list the eight rules of gereh. We can find the first attempt to formulate these rules in the paper by Gaganov (1958). However, he missed a few essential points probably because he did not have experience with some kinds of geometric patterns. Most of the terms used here will be discussed on the next page.

RULE G1

Gereh tessellation tiles are convex and symmetric.

RULE G2

Gereh tessellations are edge-to-edge.

RULE G3

Tessellation polygons are complete inside the contour or cut along their symmetry lines by the contour's edges.

RULE G4

If a single line of the pattern touches an edge of a tile, then on the other side of that edge, another line is drawn in a mirrored direction.

RULE G5

If two lines touch the edge of the polygon, they continue on the other side of the edge without changing direction or stop on the edge of the tile. Note, angles between the lines and edge of the tile are equal.

RULE G6

Two lines of the pattern inside a tessellation tile bend in places where their directions intersect, or on angle bisectors, or in other places obtained through a specific geometric construction.

RULE G7

Lines of the pattern can stop only on the edge of the contour.

RULE G8

For each tile of the tessellation, the pattern inside it has the same symmetry as the tile itself.

Some of these rules can be ignored if we have a good reason for it. In particular ignoring rule 8, we may get some interesting results.

Some necessary explanations

Most of the terms in rules **G1**, ..., **G8** use mathematical language that sometimes may need a few words of explanation.

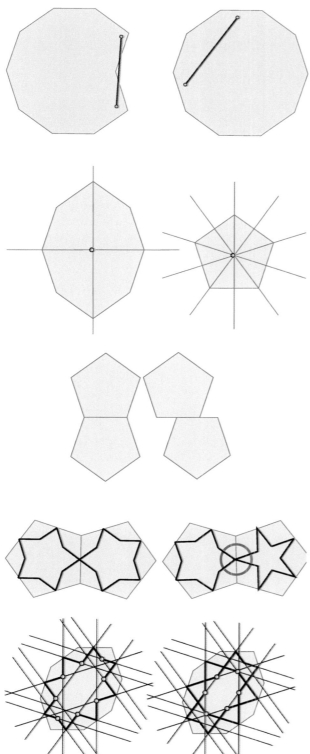

Fig. 1.7. (RULE G1) The polygon on the right is convex. Any two points inside of it can be connected with a segment that is entirely inside the polygon.

The polygon on the left side is not convex as it has a notch, and we can find a straight line connecting two points inside the polygon and crossing its edge.

Fig. 1.8. (RULE G2) The left polygon, a decagonal shield, has two symmetry lines; we call them mirrors. There are no other symmetry lines in this shape. The pentagon has 5 symmetry lines, and each of them can be used to make a mirror reflection through it.

We say that the left polygon has D2 symmetry and the pentagon D5. We can also consider symmetries of a figure in respect of rotations about its center. Thus the left figure is C2, which means that we have to rotate it 90 deg. twice around the center to get back. Pentagon has C5 rotational symmetry.

Fig. 1.9. (RULE G2) The two pentagons on the left are edge to edge. This means they have a common edge. The two pentagons on the right are not edge to edge. Their edges only partially overlap. From the designing point of view, this second situation may be a real problem. Thus we never ignore this rule.

Fig. 1.10. (RULE G5) In the left image, lines of the pattern in both pentagons continue without bending. In the right image, lines of pattern in both pentagons do not match each other. Thus the place encircled looks bad. We should avoid such situations. However, sometimes breaking rule G5 gives an interesting result.

Fig. 1.11. (RULE G6) Here we show two of the possible situations. One can finish this pattern in places where directions of lines intersect for the first time. This way, we may get the pattern and a large empty space inside (left image).

We can also continue along direction lines and produce an additional internal part. In both cases, we will create a valid pattern. This concept is essential when we deal with big stars.

Rules **G1**, **G2**, and **G7** should never be broken. The remaining rules can be ignored sometimes if we have a good reason for it.

Drawing convention

Drawings and geometric constructions are the major part of this book. To make them easier to read, we will use the following conventions:

- Red lines will be used for mirror symmetry lines. In fact, we are rarely forced to make a mirror reflection of anything. Usually, mirror images of any element of a pattern come naturally by extending appropriate segments.
- Thin, black, solid lines will denote construction lines. Fragments of a pattern will usually be located along such lines.
- All thin lines will be wiped out when the design of a template or a pattern is finished. For this reason, I suggest using a soft pencil for drawing them. It will be easier to erase them at the end of the drawing process.
- Thick lines will be used for the contour and the pattern lines. It would be wise to use a hard pencil or even drawing with a fine line pen or a gel ink pen.
- The symbol ■ will be used to indicate 90 degrees angles. It will show where two lines are perpendicular.
- Symbols '|' and '||' will be used to indicate that two segments have the same length.
- Sometimes to make our drawing clear, the tessellation polygons will be colored using different shades of the same color.
- Points will be shown as small circles. To avoid crowded drawings, we will show only those points relevant at a given stage of construction.

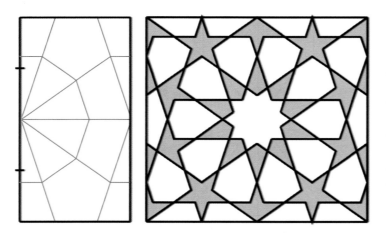

Fig. 1.12. Two stages of a pattern construction

The left figure shows tessellation ready to be filled with decoration. Note – a regular polygon can be treated as a whole or as a collection of long triangles. From a practical point of view, drawing all the symmetry lines of a polygon can sometimes be very useful. Here we show only selected symmetry lines of the decagon. We do not display all the symmetry lines of all polygons. Note also – edges of a tile are local symmetry lines. Thus we draw edges using thin red lines.

The right figure shows a finished pattern. The coloring may be more sophisticated or none. This is not important from a pattern designing point of view.

I guess we are ready to do our first working example and try the gereh theory in action.

2. Selected geometric constructions

This chapter will discuss a few geometric constructions that could be useful for designing patterns in this book. However, if you do not want to complicate your life, you can simply use a selection of drawing tools available on the market. You can buy a ruler, compasses, drawing triangles, and a good protractor. That is all you may need.

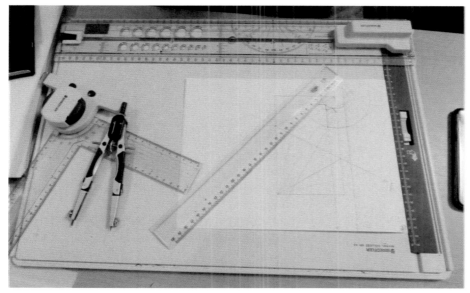

A professional drawing board can be useful in designing complex patterns.

Depending on your pocket's content, you can buy very simple tools or fancy and expensive ones. If you want, you can invest in a professional drawing board like the one shown here. It is up to you.

One can also use a computer program to make geometric constructions and draw geometric patterns of any type. Every drawing in this book was made using Geometer's Sketchpad® a very simple computer program for teaching geometry in schools worldwide. GSP, once proprietary software, is now free and available for anybody who wants to use it. It is the only geometry program I know that can handle enormous data for complex geometric patterns. A brief tutorial of Geometric Sketchpad is enclosed at the end of this chapter. Each of the geometric constructions presented here can also be done with GeoGebra that is essentially very similar to the Geometer's Sketchpad (check www.geogebra.org/)

Now, let us get back to geometry.

In this chapter, capital letters A, B, C... will denote points. We will use a small circle to show where the point is located. Small letters $a, b, c, ...$ will represent lines. Segments will be described by their endpoints. Thus AB is a segment with ending points A and B, while |AB| means the length of segment AB. Lines do not have ends. However, sometimes it will be convenient to use the notation 'line AB', meaning that we deal with the line passing through the points A and B. Circles will always be labeled as follows c_1, c_2, c_3,.... The circle with center in A passing through the point B we will represent by the formula $c(A, AB)$.

1. Constructing a perpendicular line to a given line or segment

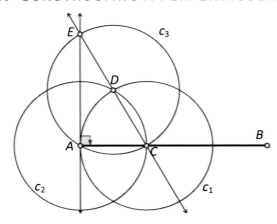

Fig. 2.1. Draw segment AB, and select any point C on it. Then draw the circles $c_1=c(C, AC)$, $c_2=c(A, AC)$ and $c_3=c(D, DA)$, in the order shown here by their labels. After drawing a line through points C and D, we get point E. The line through A and E is the perpendicular line to AB we are looking for.

2. Dividing an angle into two equal parts

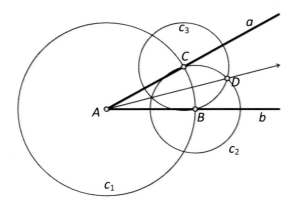

Fig. 2.2. Start with the angle between lines *a* and *b*. Select any point B on one of them. Then draw three circles $c_1=c(A, AB)$, $c_2=c(B, BC)$ and $c_3=c(C, CB)$. By drawing a line through points D and A we divide the angle into two equal parts.

Important – there is no general construction for the trisection of an angle. However, some angles can be easily divided into three equal parts, e.g., angle 90°.

3. Dividing a segment into two equal parts

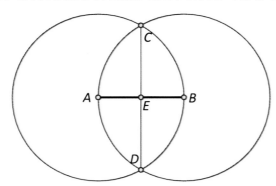

Fig. 2.3. Segment AB is given. Draw two circles $c(A,AB)$ and $c(B,AB)$. Connect their intersection points C and D by a line or a segment. Point E is the center of AB.

4. Drawing a line parallel to a given line and passing through a given point

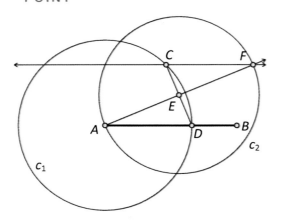

Fig. 2.4. Segment AB and point C are given. Draw circle $c_1=c(A,AC)$. You will obtain point D on AB (or line passing through AB). Find the midpoint E of the segment CD. Draw circle $c_2=c(E,AE)$. The line passing through points C and F is the line parallel to segment AB.

14 | Practical geometric pattern design – decagonal patterns in Persian traditional art

5. Constructing angles 18, 36, and 54 degrees

These three angles are essential. We will use them frequently.

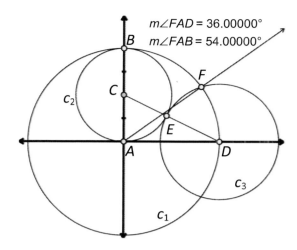

Fig. 2.5. Start with two perpendicular lines passing through the common point A. On one of them, select any point B.
Draw circle $c_1=c(A, AB)$. Divide segment AB into two equal parts. Then draw the two remaining circles in the order they were labeled. The line passing through A and F forms a 36 degrees angle with AD. The other angle is 54 degrees. The 18 degrees angle we will obtain by dividing the angle 36 degrees into two equal parts.

6. Dividing a right angle into 5 equal parts

This construction is a basis for all patterns in this book. It can be worth drawing it on paper, copying only the resulting lines on tracing paper, and then later using the tracing paper template whenever we need it. A more universal template was enclosed at the end of this chapter.

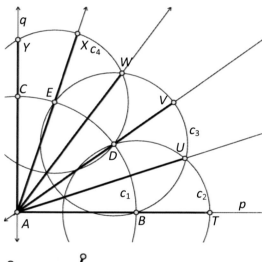

Fig. 2.6a. Start with two perpendicular lines p and q. On the horizontal line, select any point B. Construct angle 36 degrees exactly as we did in the previous construction. Draw a large circle $c_1=c(A,AB)$. Then draw the three remaining smaller circles in the order shown by their labels. Finally, connect points Y, X, W, V, U, and T with point A. Segments AT, AU, AV, AW, AX, and AY form our template. Copy only them on tracing paper.

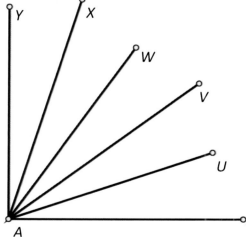

Fig. 2.6b. This is the final view of the template for dividing the right angle into 5 equal parts. Note, each part will have 90/5=18 degrees. Then we can easily calculate:

$$\angle TAU = 18°$$
$$\angle TAV = 36°$$
$$\angle TAW = 54°$$
$$\angle TAX = 72°$$
$$\angle TAY = 90°$$

7. Constructing the golden section

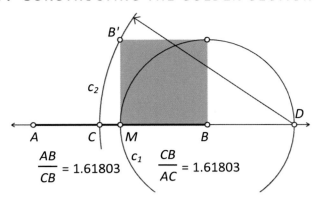

Fig. 2.7. On a few occasions, it will be useful to divide a segment AB into two parts AC and CB, such that |AB|/|CB|=1.61803... and |CB|/|AC|=1.61803...that we know as a golden ratio and sometimes denote it using the Greek letter Φ. It is a handy construction in creating some complex patterns.

Find midpoint M of the segment AB and create a square with side MB. Expand segment AB to the right and draw the two circles shown here. Point C is the golden section of AB.

By dividing segment AB into two parts AC and CB, such that |AB|/|CB| = Φ, we obtained two specific segments – a shorter one AC and a longer CB. In most of the constructions of decagonal patterns, we will frequently be meeting two such segments having this specific proportion. We will usually denote them as S and L, where S means short, and L means long. As we will see later, most of the figures used in decagonal patterns will have sides that are combinations of S and L. In more advanced topics, we may have a third value: $X = L + S$. Here X is the shortcut from XL. Note, the properties of these three values according to the golden ratio definition give us a very useful formula.

$$\frac{X}{L} = \frac{L}{S} = \Phi = 1.61803 \ldots$$

Note also, although, in the next few chapters, we will deal mostly with tessellations where edges of tiles are L and S in size. In many complex patterns, we may deal with all three sizes in one design.

8. Dividing a segment into a given number of equal parts

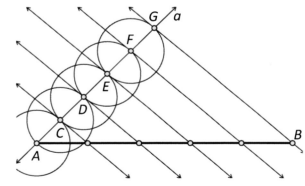

Fig. 2.8. Suppose that we want to divide segment AB into 5 equal parts. We draw any slant line passing through point A. Then on this line, we select any point C. By drawing 5 circles with the same radii equal to AC we obtain points C, D, E, F, G.

Draw a line connecting points G and B. Then draw four more lines parallel to line GB and passing through points F, E, D, and C. The intersections of these lines with segment AB divide it into 5 equal parts.

9. Dividing a given segment into three parts with proportions L, S, L

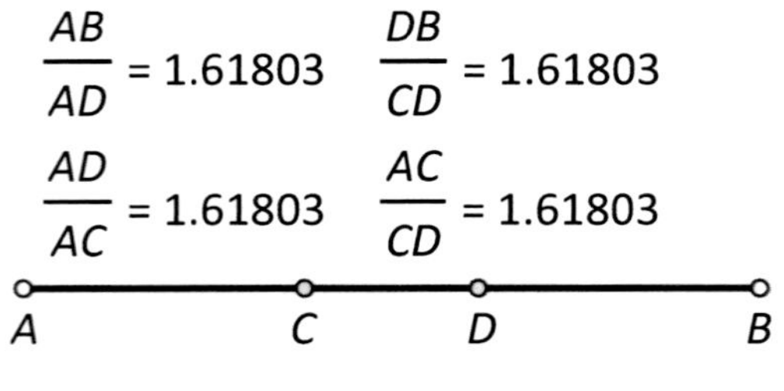

$$\frac{AB}{AD} = 1.61803 \qquad \frac{DB}{CD} = 1.61803$$

$$\frac{AD}{AC} = 1.61803 \qquad \frac{AC}{CD} = 1.61803$$

Fig. 2.9. Segment AB is given. Construct point C that is the golden section point for AB, and then construct the point D that is the golden section for CB.

10. Construct a pentagon with a given side

The construction of a regular pentagon will be very useful in more advanced topics. However, it is convenient to have it together with all other constructions.

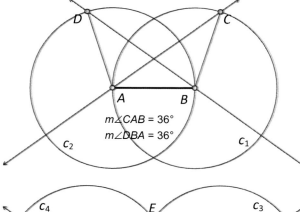

Fig. 2.10a. Segment AB is given. Draw two lines, one passing through A, and one through B, such that the angle between each of them and AB is equal to 36 degrees. For this purpose, we can use the template created in drawing 6b.

Then draw two circles with centers A and B respectively and radii equal to AB.

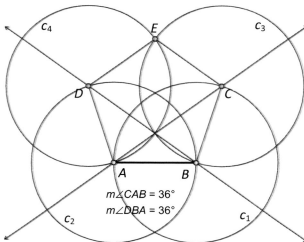

Fig. 2.10b. Draw two more circles with centers C and D and radii equal to AD and CB. Their intersection is point E. Connect points B, C, E, D, and A to get a regular pentagon.

11. Drawing a regular decagon with a given side

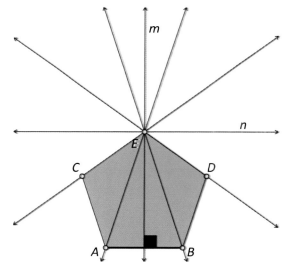

Fig. 2.11a. Construct a pentagon with a given side and with vertices labeled as on the picture. Construct a line that passes through vertex E and is perpendicular to AB. This will be the line m. Then construct another line perpendicular to m passing again through vertex E.

Draw lines CE, AE, BE, DE.

2. Selected geometric constructions | 17

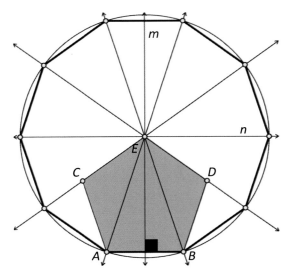

Fig. 2.11b. Now, we need a circle with the center in E and passing through point A. Points of intersection of the circle with the lines passing through E form a regular decagon with given side AB.

12. Constructing the mirror reflection of an object about a line or a segment

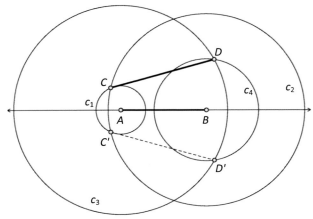

Fig. 2.12. The line passing through points A and B is given. The segment CD should be reflected about line AB. Draw two circles $c_1=c(A,AC)$ and $c_2=c(B,BC)$.

Their point of intersection is C', a mirror reflection of point C.

In the same way, by drawing two circles c_3 and c_4, we produce point D' that is a mirror reflection of point D. Segment C'D' is a mirror reflection of the segment CD.

While drawing a pattern by hand to create the mirror reflection of anything, it is enough to copy the drawing onto tracing paper, put the tracing paper copy upside down, and position it correctly.

Useful template

The drawing shown below can be used as a template for many constructions in this book. Copy it on tracing paper and keep it for further use.

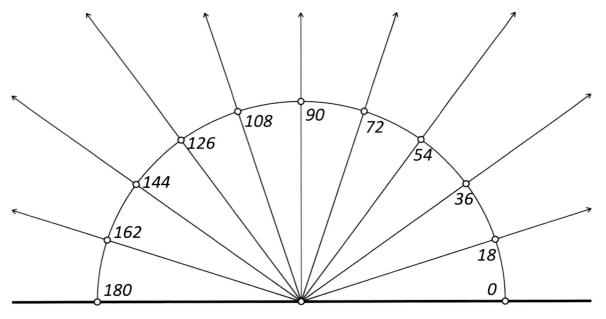

18 | Practical geometric pattern design – decagonal patterns in Persian traditional art

➢ Using Geometer's Sketchpad in pattern design

Geometer's Sketchpad mimics all traditional drawing techniques with compasses and rulers. We can mark points; draw segments, lines, polygons, etc. Many basic constructions are available through Sketchpad menus. This way, we can make our drawings more accurate. For example, while drawing geometric patterns, we often draw perpendicular or parallel lines. Drawing them by hand leads to various inaccuracies and design problems. Operations 'Perpendicular line' and 'Parallel line' from the 'Construct' menu produce exact results. This is a significant help.

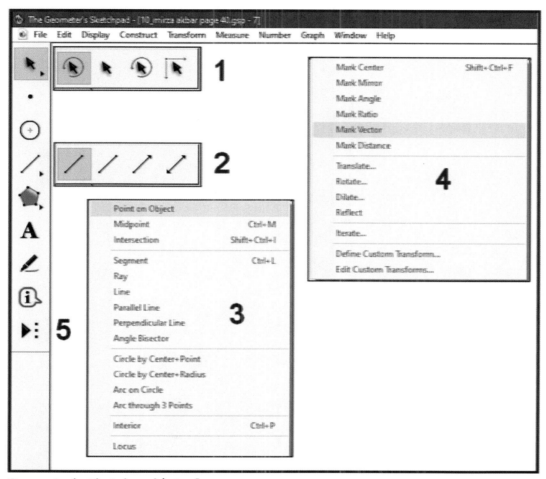

Geometer's Sketchpad interface
- Operations for manual translating, rotating, and resizing objects (see nr. 1 in the image)
- Tools for drawing segments, rays, and lines (see nr. 2 in the image)
- Tools for performing basic constructions (see nr. 3 in the image)
- Tools for performing transformations of objects – translations, rotations, dilations, and reflections (see nr. 4 in the image)
- Entry to the user-made custom tools (see nr. 5 in the image)

As we can see from the above image in Sketchpad toolboxes and menus 'Construct' and 'Transform', we have all tools needed to draw geometric patterns in the same way as with compasses and ruler.

One can add to its repertory the so-called user-made custom tools. They are computer forms of working with a tracing paper. In pattern design, we often create a motif, and then we copy it with tracing paper into all locations where we need it. In Geometer's Sketchpad, we do the same with custom tools. Here is one example.

A CUSTOM TOOL TO DRAW A REGULAR PENTAGONAL STAR

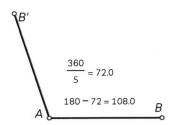

We start by drawing the segment AB. Points A and B are essential. These are points for creating the tool.

Then we perform mental or computer calculation of angles. We know that any regular pentagon can be divided into 5 identical triangles with one vertex in the center (the angle 72.0 degrees) and two other angles 108.0 degrees together. In the next step, we rotate segment AB 108.0 degrees about point A. To do this, we need to double click on point A, select segment AB with point B, and from the menu transform, use operation rotate. Insert 108 degrees as the angle of rotation. This way, we get the drawing shown here.

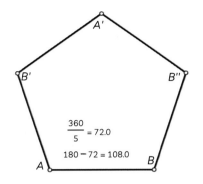

In the next step, we perform the consecutive rotations about points B' and A'. The last segment B"B, we draw without rotations.

Now we select points A and B (in this order) and then all remaining points and all segments. It is more convenient to select multiple objects with the [Shift] key pressed down.

To complete the creation of the tool to draw a pentagon, we select from the custom tool menu option 'Create a new tool' we give it a name, e.g., 'pentagon'. Now the tool is ready. With it, we can draw a few pentagons to check how it works (see below).

Three pentagons were created with the custom tool 'pentagon'. We started with the central pentagon, and then we added the two pentagons on the left and right sides.

Now, we can use the 'pentagon' tool to create a new tool. Let us call it 'pentagonal star'.

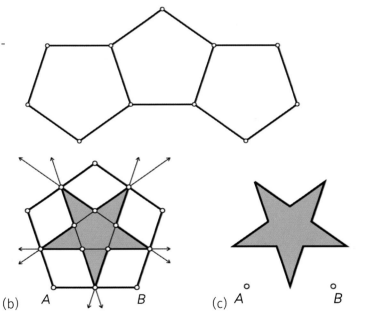

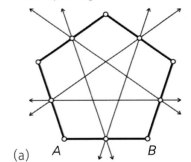

Creation of the 'pentagonal star' tool

We start by drawing a pentagon (use the custom tool). Then we create midpoints of each side of the pentagon, draw the lines shown here (a) and (b). In the final stage, we hide all that is not present in the drawing (c). Now again, select points A and B (in this order) and the remaining elements shown in (c). Again using menu custom tools, we create a new tool, 'pentagonal star'.

3. Styles of decagonal patterns

This is the right time to explain what we mean by the term 'decagonal pattern'. As we have seen in chapter 1, the tessellation in the example from the Topkapi scroll was in some way derived from a regular decagon. We had tiles that were part of a decagon. We also had two pentagonal shapes that can be constructed from a regular decagon. Then we had other shapes that, as we will see later, can be produced from a regular decagon or a regular pentagon. Thus, every pattern with a tessellation based on a regular decagon's geometry we will call a decagonal pattern.

In this chapter and in a few next chapters, we will introduce several types of decagonal patterns with Persian origins. For each of them, we will use a different term to describe the given type. I am not trying to make any formal classification of decagonal or any other type of patterns. The terms used in this book should be understood as labels that we attach to a particular group of patterns to distinguish them from another group. It works in the same way as when we compare objects in real life – we often say that a given thing looks like another given thing. A good example is the sentence, "my car looks like the car of my neighbor; it is blue with scratches on one side."

In the past, there were some attempts to classify geometric patterns with given local symmetries. For example, Anthony Lee tried to establish a classification of decagonal patterns (Lee, 1975). In this chapter, we will come back to his classification, or let us say typology, and compare it with our observations.

As we will see later, there is no easy way to make a clear classification of geometric patterns inside the group of patterns with decagonal local symmetries. Persian and later Ottoman artists created designs mixing different features together. Thus while visiting mosques in Turkey, we may see designs combining Persian and Ottoman features. There are also numerous gerehs with some non-typical geometric solutions. Two collections of such designs can be seen in Istanbul in Rustem Pasha Mosque and in Sultan Selim Mosque. Another splendid gereh on the main gate to the Topkapi palace's external park is often considered incorrect. In several places in the Middle East, we can see designs with some unusual features. A few of them we will discuss in this book.

Several unusual designs we will investigate in the another book devoted to Seljuk and Ottoman designs.

➢ Project 3.1 – Our first contour and tessellation

We will start by creating a typical contour for decagonal patterns. We will construct a tessellation based on this contour, and finally, we will create a few patterns that can be constructed using this particular contour and this particular tessellation.

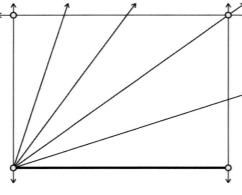

Fig. 3.1a. Start by drawing a horizontal segment. This will be the bottom line of the contour. At both ends of this segment, construct two lines perpendicular to it.

Divide one of the right angles into 5 equal parts.

Mark the point where the second section line from the bottom intersects with the right vertical line. Draw a line passing through this point and parallel to the bottom segment. The rectangle bounded by these lines is our first and the most popular contour for many decagonal patterns.

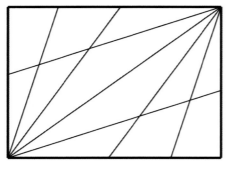

Fig. 3.1b. Draw edges of the contour, and divide the top-right angle into 5 equal parts. This way, we will get the network shown here. Note, the whole network is based on the division of the right angle into 5 equal parts. These are the white lines on the photographs from Rempel's book. These look like strings or ropes tied on a frame.

Now it is time to start tying knots.

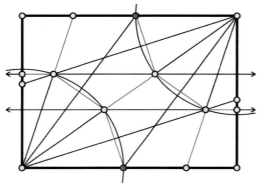

Fig. 3.1c. Use the two red points on the contour's edges to draw two circles with centers in opposite corners of the contour and passing through the red points. Then make knots where the 'strings' and circles intersect. All knots are marked here as small circles. Finally, connect these knots with red segments. This is our complete tessellation. Note how the short red segments near the right and left edge were created. Now we can remove all construction lines and leave only the most necessary elements, i.e., the network of red lines and the contour.

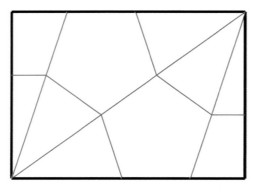

Fig. 3.1d. Here we show the final view of the tessellation created in this example. We have here two regular pentagons. So, there was no need to construct them separately. We have here two-quarters of two decagons located in the bottom-left and top-right corner. We divided them into long triangles. Finally, we have here two-quarters of trapeziums located in the top-left and bottom-right corners.

If we denote lengths of the sides of decagons and pentagons as S, then the long sides of triangles will have length L, where L/S= 1.61803=Φ (golden ratio). Finally, the trapeziums will have short sides S, and one long side L. Thus we are still dealing with the golden ratio.

As I mentioned before, the tessellation from this first project is the most popular tessellation for decagonal patterns. Everything depends on how the first line of the future pattern will be placed. Here is one such pattern.

➢ Project 3.2 – The Nodir Devon Madrasah style

Fig. 3.2 A ceramic mosaic from the inner portal of the Nodir Devon Madrasah from Bukhara.

The goal of this project will be to reconstruct the pattern shown in the photograph.

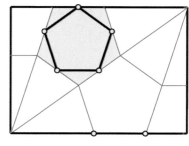

Fig. 3.2a. Take one of the pentagons in our tessellation and mark the center of each of its sides. Now, draw segments connecting the centers of its adjoining sides. This way, we will get a nice smaller pentagon. We can do the same for the other red pentagon. This is your task.

Fig. 3.2b. After filling the two red pentagons with the pattern we extend the lines of this pattern onto the left quarter of the decagon. By following the thin lines, we make a few more knots, and this way, we get the pattern inside the decagon. We do precisely the same for the other quarter of decagon and halves of trapeziums.
The template for the Nodir Devon Madrasah pattern will be ready.

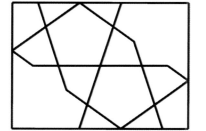

Fig. 3.2c. Finished template for the Nodir Devon Madrasah pattern. Although this is the simplest pattern that could be designed on the tessellation from project 1, we will see later that this template can be used to create several more complex patterns.

The style of the pattern shown in the photo (figure 3.2) we will call the Nodir Devon style. There is no particular reason to give it this or any other name. We just need a name that can be used for future reference. We may use a more or less descriptive name for this style. Tony Lee, in his manuscript (Lee, 1975), refers to it as Type I. However, some of his later types are just variations of this one. We can refer to this particular style by the measure of angles used in it. But this can be a bit misleading – which angles should we use? The same pattern can be seen in many other places, e.g. in the Karatay Madrasah in Konya, the Tilia Kari Madrasah, and Bibi Khanum Madrasah in Samarkand. Thus we could use another name to identify this style. It does not matter what name we use as long as we know what kind of pattern we mean.

➢ Project 3.3 – The Kukeldash Madrasah style

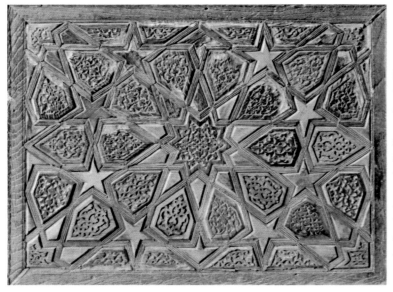

Fig. 3.3 The Kukeldash Madrasah design

The photo shows the bottom part of the doors to the Kukeldash Madrasah in Bukhara, in Uzbekistan. There are two designs on these doors. The other one, more complicated we will discuss it later.

The Kukeldash Madrasah in Bukhara was built in (1568-1569), and the doors to it are the original doors made around 1569-70. Thus, this is probably one of the oldest wooden artifacts with such a design.

This is also one of the oldest known examples of kundekari art.

Let us start with the same tessellation. This time we will place our first line in a slightly different way. We will see how the pattern changes depending on how the first line of the pattern is selected.

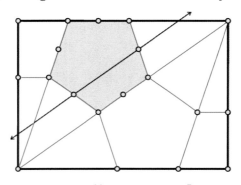

Fig. 3.3a. We start constructing the pattern inside a pentagon by drawing the first line through the two points shown in the figure. Of course, now, due to the symmetry rule G8, we have also to draw four similar lines passing through the other pairs of midpoints on the edges of the pentagon.

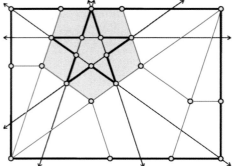

Fig. 3.3b. After drawing the other construction lines, we get a nice star with sharp ends. Its center can be filled with additional segments forming there a small pentagon. This way, we will get a slight variation of this gereh.

Now let us draw identical lines in the other pentagon, and we will get almost everything that we need to construct the complete template.

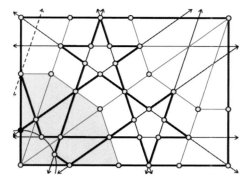

Fig. 3.3c.d.e. Here we have both pentagonal stars finished and their edges extended into the left-bottom and right-top areas of the contour.

Most of the pattern in the quarter of the decagon is a natural extension of the pattern from the pentagons. Note – how the black point on the left edge of the contour was used to make the inner part of the pattern inside the decagon.

Finally, we have to draw the pattern in the other quarter of the decagon and in the left-top and bottom-right areas. For this reason, I created here the dashed line that is parallel to one of the edges of the top pentagon.

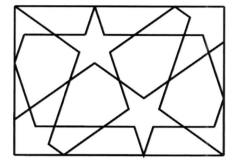

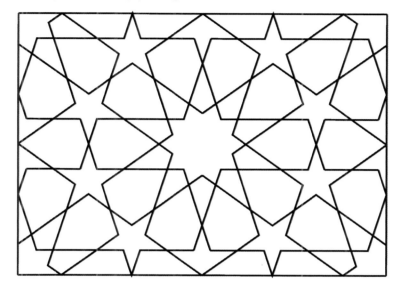

Above we show the finished template and right a pattern made from it.

In the title for this project I used the term 'Kukeldash Madrasah style'. Again there is no particular reason for selecting this or any other name. We need it only for reference purposes. The pattern shown above is one of the most popular patterns in Uzbekistan, Iran, and Turkey. We can find it in various versions all over the Middle East. Its specific feature are the sharp angles in pentagonal stars.

We will talk about other Kukeldash Madrasah style patterns later. Tony Lee refers to this style as type II.

➤ Project 3.4 – The Persian style

Fig 3.4 This particular photo was taken in Isfahan Jame Mosque. It shows a fragment of a large ceramic mosaic on the wall. Patterns like this one are very popular all over Central Asia. In Uzbekistan, we may see it on the wall of one of the madrasahs in Regestan square in Samarkand and many other places. A typical feature of this design is that the pentagonal stars have parallel, or almost parallel, edges. This, of course, will influence all other shapes used in this design.

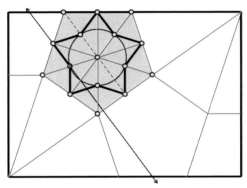

Fig. 3.4a. We start again with the same tessellation and midpoints on the sides of one of the pentagons. We draw all symmetry lines of the pentagon, and then from one of the sides, we construct a line passing through the center of the side and parallel to the dashed symmetry line. This is our first line.

We use the intersection point of the first line with one of the symmetry lines to draw the circle shown here. This way, we got a framework to draw a star pattern inside the pentagon.

We should do the same with the other pentagon.

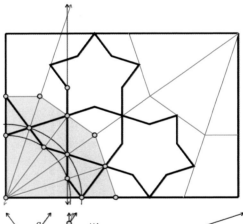

Fig. 3.4b. The pentagonal stars are ready. Now, we extend one of the edges of such a star. Just one is enough. We get a vertical line crossing some of the symmetry lines of the decagon in the left-bottom corner. Now we use the intersection points of the vertical line with decagon symmetry lines to create the two circles. This is our framework for drawing the pattern inside the decagon.

We should repeat the same in the right-top corner.

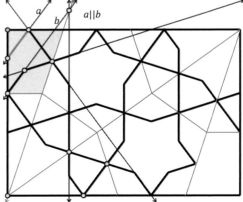

Fig. 3.4c.d.e. Here I show how we could proceed with the finishing touches in half of the trapezium. Again we extend some lines from the pattern in the pentagon. Lines *a* and *b* in the left-top corner are parallel.

Note – this pattern may or not have the thick gray segment connecting two points in the top-left corner. In the pattern from Samarkand, we do not have it, but in many other places, it is included.

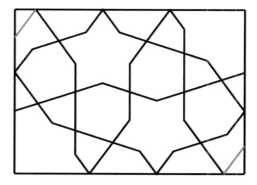

Above we show the finished template and to the right a pattern made from it. Here we show the version with the optional gray segments.

At the beginning of this project, I used the term 'Persian style.' Tony Lee refers to it as Type III. Let us look for a while at specific features of these three styles or types.

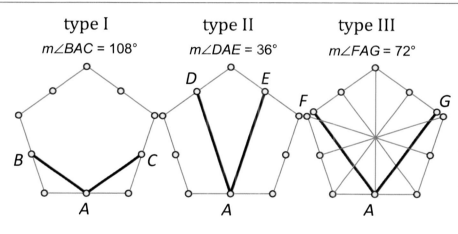

type I
$m\angle BAC = 108°$

type II
$m\angle DAE = 36°$

type III
$m\angle FAG = 72°$

Types of decagonal patterns according to Tony Lee

Type I has the widest angle, Type II has the narrower angle, and finally, Type III has a medium angle. The first type we called the Nodir Devon Madrasah style, the second one as the Kukeldash Madrasah style and the third one as the Persian style.

➢ Project 3.5 – The Tan Sahid Mosque style

This is not the most popular kind of pattern. However, we see it in a few places in Central Asia and Egypt, e.g., the Tan Sahid Mosque in Iran, Madrasa Amir Sungur Sa'di in Cairo, and Turkey, Sunqur Bek mosque. A very beautiful pattern using this style can be seen in the Rustem Pasa Mosque in Istanbul.

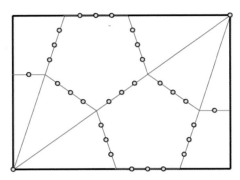

Fig. 3.5a. By dividing each side of each pentagon into four equal parts, we get our template in a modified form. This time the centers of the sides will not be used. These are just a reminder of the previous projects. We will work mostly with the new points.

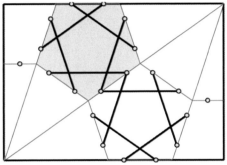

Fig. 3.5b. Here I show what we can do with these new points. The midpoints of the sides of the pentagons are hidden.

If we divide the edge of a pentagon into three equal parts, then the rhombus on the common edge of two pentagons will get smaller, and the pentagons with black edges will get larger. The distance between points shown here can be larger or smaller. Each time we will get a slight variation of the same pattern. This issue we discuss later this chapter and show on the drawing 3.6e.

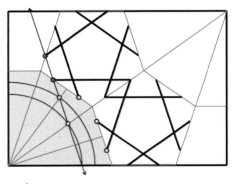

Fig. 3.5c. The line passing through the two red points will allow us to create a pattern inside the bottom-left corner's decagon. After drawing the two circles passing through the intersection points of this line with the decagon's symmetry lines, we have a framework ready to draw the pattern.

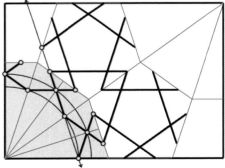

Fig. 3.5d. Here I show how the pattern inside the decagon was created.

This part may have some interesting variations. One can consider this pattern as a structure of some specific overlapping kites. Interestingly, the lengths of sides of these kites have a proportion $L/S = 1.61803 = \Phi$. Thus the golden ratio is still with us.

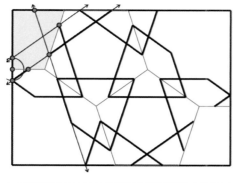

Fig. 3.5e.f.g. To fill the corner with a pattern, we need to extend some pattern elements from the decagon and pentagon. We get a framework that will allow us to complete another kite in the top-left corner again.

Below we show a complete template (left) as well as a pattern created with this template. Here we see the nice structure of the ring of ten kites located around the pattern's center.

Bourgoin (1973), shows similar patterns for other types of symmetry – dodecagonal and octagonal.

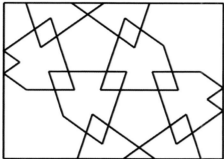

Gereh from this project is in many aspects different than those from previous projects. We can easily distinguish its style from many other patterns. In Tony Lee's notes, this pattern is labeled as Type VI. Unfortunately, his types IV and V are just modifications of Type I. In the future, we will refer to it as the Tan Sahid Mosque style or simply the Tan Sahid style. This pattern may have a few interesting modifications. Here is one of them.

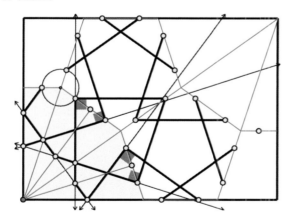

Fig. 3.5h. Sometimes breaking the rules gives us a chance to produce something unusual. Here – note – we draw lines perpendicular to the lines of the pattern inside the two pentagons. The squares show how we did it. Now, all we have to do is to draw the pattern along these perpendicular lines. This way, we will get a pattern with modified kites.

Fig. 3.5i.j. Template and a pattern made using four copies of this template.

➤ Project 3.6 – Kites and roses

In gereh from project 3.5, we have a large empty space inside. Such space can be a bit dull if we develop this pattern on wooden doors. For this reason, some of the woodworking masters fill it with a nice rosette. This is the purpose of this project.

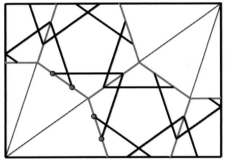

Fig. 3.6a. We draw the pattern inside pentagons exactly the same way as in project 5. But then we again break the G8 rule. This time we divide edges of each pentagon into three equal parts.

We extend one of the lines from the pattern in the pentagon, draw a circle as shown here, and then use the red point to draw two lines parallel to the long triangle's edges. This will be a starting point to make a nice rosette inside the decagon.

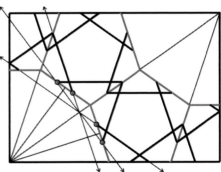

Fig. 3.6b. Here we have a complete pattern inside the long triangle. We have to draw an identical pattern in the remaining triangles in this decagon and the one on the top-right of this image. The easiest way is to copy it to the other places using tracing paper.

Note that here we broke rule G4. Lines of the pattern from the pentagon pass the edge of the tile without changing their direction. According to rule G4, they should bend here in a mirrored direction.

In the next figure, we see a complete template and a pattern made from it.

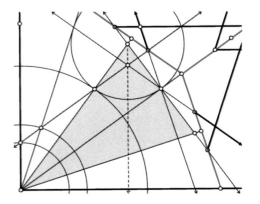
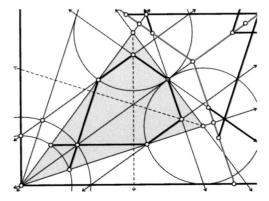

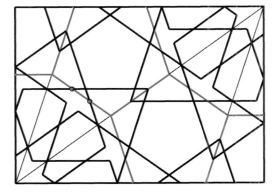

Fig. 3.6c.d. Template and a pattern made using four copies of this template.

A pattern identical to the one we created here can be found in the book by D'Avennes (2007). The Miriam and Ira D. Wallach Division of Art and Architecture in New York Public Library shows it with French inscription 'Art arabe : maison de Sidi Youçouf : porte intérieure (XVIIIe. Siècle). This means Arab Art: House of Sidi Yusuf: Interior Door (Eighteenth Century). Bourgin (1973) shows this pattern in his plate 179.

This pattern brings us to an interesting discussion. Note – while developing the pattern in pentagons, we decided to split the pentagon's edge into three parts with lengths 1/3, 1/3, and 1/3 of the edge of the pentagon. This was our arbitrary decision, and one could split the pentagon's edge into three parts with different proportions. In fact, medieval artists often used different proportions. The resulting patterns were similar but not identical. Let us see what is possible.

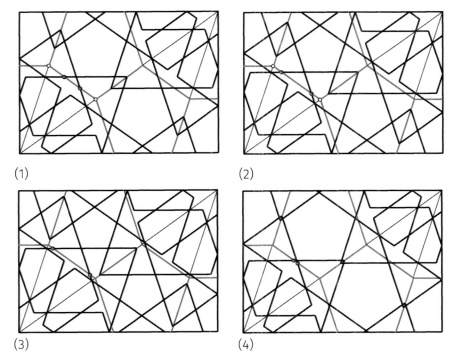

Fig. 3.6e. Different versions of the pattern from project 5 depending on how we divided the edge of the pentagon. Here we have:

(1) Original: 1/3, 1/3, 1/3

(2) equal lengths 1/6, 4/6, 1/6

(3) Wide center 1/10, 8/10, and 1/10

(4) Narrow center: 5/12, 2/12, 5/12

In each case, the rosettes in the corners may look quite different, and the other shapes also.

The situation described here was observed by other authors. In his book (Wichmann, 2018), Brian Wichmann devoted large section to patterns with fixed and variable diamonds.

➢ Project 3.7 – Kites and roses 2

This is rather unusual and, at the same time, a rather difficult pattern to construct. If you feel that it is too hard for you, you may skip it in this book's first reading. I found this pattern on a photograph from one of the Central Asian scrolls. After trying many ways, the golden section proved useful in reconstructing this pattern. So, let us start.

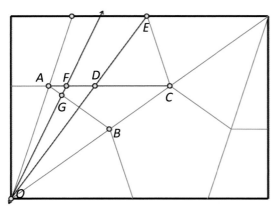

Fig. 3.7a. Construction with the **golden section**

We start with the same tessellation as in all previous projects. Then we draw segments OE and AC. The distance between A and D we divide using the **golden section** construction. Point F is the result of the division. By drawing the line through F and O, we produce point G. This point will be the key to the rest of the construction.

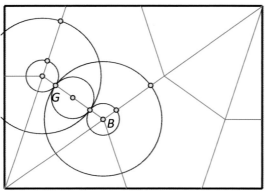

Fig. 7b. Now we have to transfer point G onto the other sides of the pentagon. Circles with centers at vertices of the pentagon can do this job for us. Here I show only the beginning of this process. You will have to do the same for the two remaining sides of the pentagon.

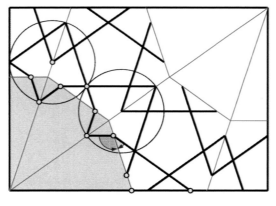

Fig. 7c. As in a few previous projects, we fill pentagons with the crossing lines. We can also do the same for the trapezium halves in both corners of the contour.

In this picture, I show how to use two circles to expand the pattern into the bottom-left decagon.

By the way, here we are, breaking rule G4. According to the rule the two angles shown here should be equal, but they are not. However, what is nice – the two segments meeting on the edge of the decagon are equal. This is the result of applying the golden section.

Fig. 7d. We extend one of the segments inside the decagon. This way, we obtain the two additional points shown in the picture. By drawing the three arcs shown here, we will make the rest of the pattern filling the decagon.

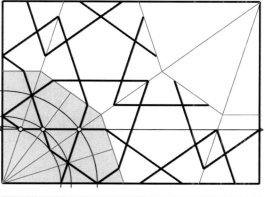

Fig. 7e.f.g. On the left, we see how we could construct the pattern inside the decagon.

Below on the left, we have a template for this pattern, and a whole pattern created using four copies of the template. The rosette in the middle is similar to the one in project 4 – the Iranian style, or if you prefer notation by Tony Lee, it is Type 3. Here we have a mix of the Persian style (rosette) and the Tan Sahid style (kites and pentagons).

This pattern comes from the Al-Ghiyasiyya madrasah, Khargird, Iran (1447).

For centuries tiling and kundekari masters in Central Asia developed very specific grids that can be used for covering windows. Such grids are usually called panjara (Persian), mashrabiya (Arabic), or şebeke (Turkish). In the next two projects, we will create such mashrabiya using the same tessellation with two pentagons.

➤ Project 3.8 – A decagonal pandjara

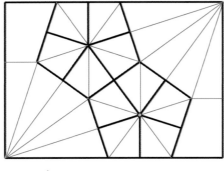

Fig. 3.8a. We adjust the tessellation by adding all mirror lines for pentagons and decagons. Then we use them to draw a pattern inside the pentagons. We also can treat the tessellation edges as a part of the pattern.

Fig. 3.8b.c. (below left). Here we show how the pattern was expanded inside the decagon. This part may have many different versions.

A very similar panjara like the one in this project can be found in many mosques in Istanbul.

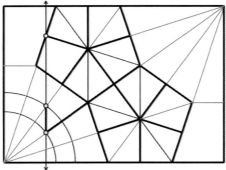

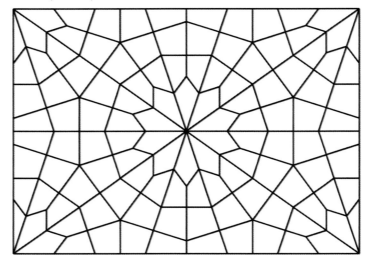

➤ Project 3.9 – A medallion from the Şehzade Mosque

In many mosques, we can find round patterns, often decagonal, on the side of a minbar or painted somewhere on the ceiling. The goal of this project will be to explore a possible way of designing such medallions.

Fig. 3.9. Decagonal medallion on the side of the minbar in the Şehzade Mosque in Istanbul. This is a very similar pattern to the one that we created a while ago. However, this time only a triangular fragment of it was used to create this large medallion. In a few other mosques in Istanbul, an identical or similar medallion was placed on the minbar's side.

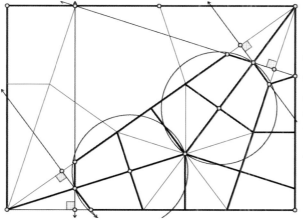

Fig. 3.9a. We start by designing only half of the template from the previous project. Note how the corners of the triangle were created. We used the two circles to mark the two points where corner stars will be created.

A triangular template like this one can be used to create a small, larger, or very large medallion. Everything depends on how many copies of it we will use. For the medallion from the Şehzade Mosque, we have to use three copies of it to create a large triangle and then use multiple copies of this triangle.

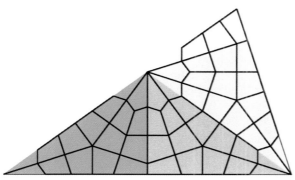

Fig. 3.9b. Here I show the second stage of constructing the large medallion.

We take three copies of the original triangular template. The original one is the darkest, then comes its mirror reflection about the right edge, and finally, with the lightest background, the third copy. Note also, we had to add a small piece (without background) to have the full triangle.

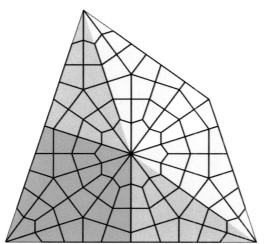

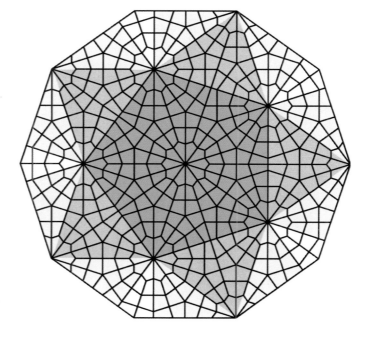

Fig. 3.9c.d. Above, we have a large kite obtained by the mirror reflection from the large triangle, and on the right, the whole medallion made from the kite by 5 rotations about its left corner point.

3. Styles of decagonal patterns | 35

Project 3.10 – The Sikandra style

In India, the decagonal pattern from Akbar's mausoleum in Sikandra is rather unusual. Some people consider it as 'incorrect.' As we will see in a moment, this pattern precisely follows the rules of gereh.

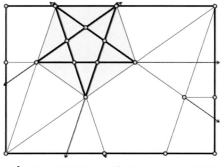

Fig. 10a. We start with the same tessellation as before. Then we draw the first line through the corners of the decagon. Due to the symmetry rule, we have to connect each pair of non-adjacent vertices with a line. Then we use these lines to create a star inside the pentagon. In the original pattern from Sikandra, this star has a small pentagon in the middle, but we can also leave the star without the pentagon.

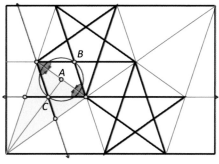

Fig. 10b. We fill the other pentagon with an identical star. Then using a circle with its center at the midpoint of a pentagon side, point A, and radius AB, we obtain point C. The lines going through it, and the vertices of the pentagon form a rhombus. We use these lines to draw the pattern inside the long triangle. Note – we are still following the gereh rules. The angles between segments on both sides of the pentagon and decagon are equal.

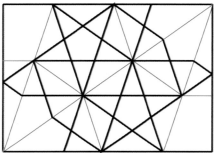

Fig. 10c. After copying the pattern from the long triangle into other places, we should get an image identical to this one here. Now we have to add the pattern to the remaining empty spaces.

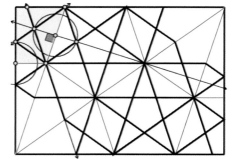

Fig. 10d. Here I show how the top-left empty space should be filled with the pattern. Again we use circles to get equal angles of the pattern on both sides of tessellation edges.

Fig. 10e.f.g. This image shows how the template for the pattern from Sikandra looks. Its original colors are similar to those that we have here. Note also – the large empty spaces in the left-bottom and the right-top corner can be filled with additional decoration. In fact, the original pattern has such a decoration.

On the next page are shown two versions of the Sikandra pattern. The left one shows what we get from our template. The right figure shows the same pattern from Birgi Aydınoglu Mehmet Bey Mosque, Odemis, in Turkey with some extra decorations.

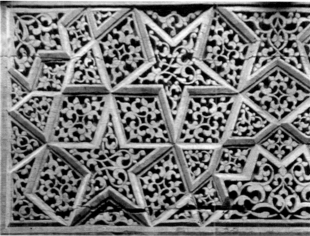

For this pattern, we used the name Sikandra style. Tony Lee calls it a Type XI pattern.

➢ Project 3.11 – Sikandra style with Mashhad twist

Patterns in Sikandra style are not very popular. However, they are extremely convenient for ceramic tiling craftsmen. They use simple shapes – mostly rhombi, triangles, and sometimes pentagons. Each of these figures uses two lengths of sides. We have again L and S with ratio L/S=1.61813... (golden ratio). As always, we can imagine that this pattern may have some slight but still interesting modifications. This is what we find in the pattern from Imam Reza's tomb, in Mashhad, in Iran.

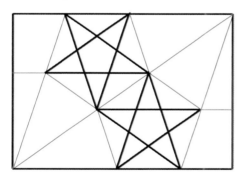

Fig. 11a. We start by drawing stars in each pentagon. This is the same as we did in the previous project. From here, we will proceed in a slightly different way. This, of course, will require breaking some rules, and specifically the rule G4.

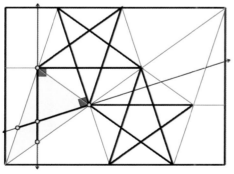

Fig. 11b. At the common vertices of a pentagon and one of the triangles, we draw perpendicular lines to the pattern segments inside the pentagon. This is something that we did previously in one of the earlier projects.

Now, we use these two new lines to draw a pattern inside the triangle. This is the only difference between this and the pattern from Sikandra.

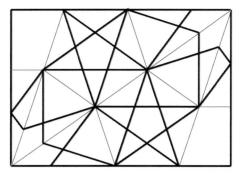

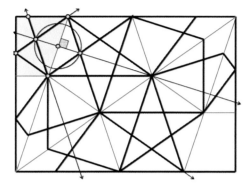

Fig. 11c.d.e. The left figure shows how the pattern inside the decagon should look. Below left, we have a solution for the left-top and right-bottom corners. Note – here, we are very consistent with what we did a moment ago. Lines of the pattern in the trapezium and decagon connected at the same vertex are perpendicular to each other.

Below right, we see a pattern made by using four copies of the template.

Note: the original pattern from Mashhad is slightly different as it uses a different tessellation. We shall come back to it in one of our future projects.

The title for this project was "Sikandra style with Mashhad twist". The major principle in constructing this pattern was to use segments or lines going out from the corners of polygons. We just ignored one of the gereh rules by using the right angles to transfer the pattern inside the decagons. Is this a different style or type? In Tony Lee's manuscript, this kind of pattern is classified as Type XII.

➤ Project 3.12 – Decagonal pattern Samarkand style

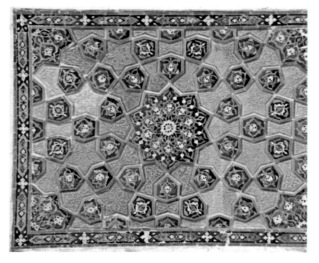

Pattern from Ulug Beg Madrasah in Samarkand

Designs like this one are often referred to as Timurid designs. But they are very typical for Samarkand and very rare outside of it. Thus I prefer to label this type of design as a Samarkand style.

This design is somewhat similar to the Persian style from project 3.4. The major difference is the size of the decagonal star and the shapes around it. These shapes are often called dolls.

A rough sketch of a very similar design can be seen in Mirza Akbar Khan scrolls.

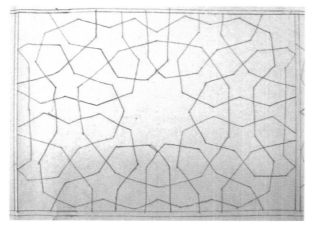

Drawing from Mirza Akbar Khan scroll

Similar to all other drawings from Mirza Akbar scrolls, the drawing is very inaccurate, but it shows the same features as the pattern from Samarkand.

The star in the center and the shapes around it, may have many variations depending on the approach we use to design this pattern.

In this project, we will show how this design can be obtained from the same tessellation that we used in all other projects in this chapter.

The drawing to the right shows how we can approach this design. We need the standard tessellation with two pentagons and two lines perpendicular to each edge of such a star. Thus we have to invent a construction of these perpendicular lines and use it to make the whole design.

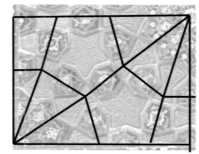

The Samarkand section

The idea for this construction comes from the previous drawing.

Start with a regular pentagon, find midpoints of each edge, draw segments connecting two midpoints on neighboring edges (CD and DE). Find midpoints of each of the two new segments (here points F and G). Through points, F and G draw lines perpendicular to the edge AB. Obtained lines are the lines needed for the construction of the Samarkand pattern.

Now we can proceed with the whole design.

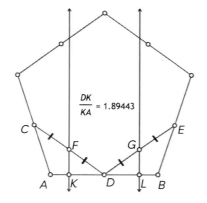

$\frac{DK}{KA} = 1.89443$

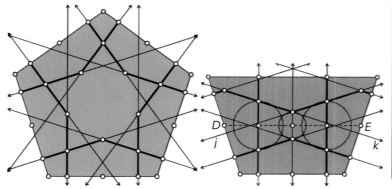

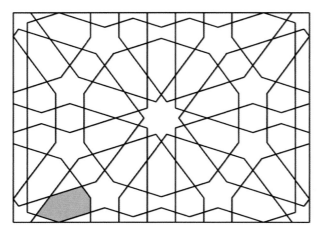

Above left
Two drawings showing how we can design patterns for pentagon and trapezium

Above right
Possible design of the pattern for the decagon

Left
Design with long hexagons. Similar long hexagons can be seen in a few places, e.g., Shamsiya Madrasah, Yazd, Iran; Ulugh Beg Madrasah, Samarkand, Uzbekistan. In the book by Mustafa Bulut (Bulut 2020), we can see it on a drawing from Sultan Han in Kayseri, Turkey.

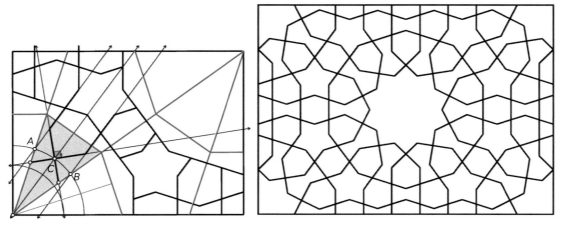

Construction of the pattern for the decagon and pattern from Samarkand
Point C is the intersection of the arc AB with the symmetry line of the shaded triangle.

➢ Project 3.13 – Multiple tessellations for one pattern

Before we close this chapter, let us look at different ways of creating the gereh from project 3.3. This will show us one important thing – the type 1 or Kukeldash Madrasah style can be created using different angles than in project 3.3. The only thing we have to change is the tessellation.

From the next series of drawings, we will learn that one pattern can be created using two completely different tessellations. We will also see how these two different tessellations can interact in one drawing.

Construction of the new tessellation is more difficult than tessellation from each of the previous projects. For this reason, we will show a very detailed construction of it. We will use the blue color to distinguish it from the tessellation from previous projects where we used the red color.

Construction of the blue tessellation

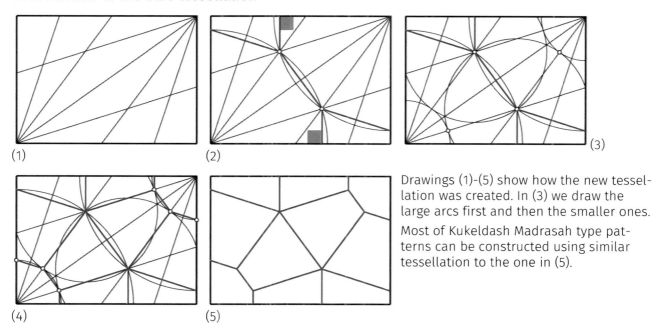

Drawings (1)-(5) show how the new tessellation was created. In (3) we draw the large arcs first and then the smaller ones.

Most of Kukeldash Madrasah type patterns can be constructed using similar tessellation to the one in (5).

Construction of the pattern using the blue tessellation

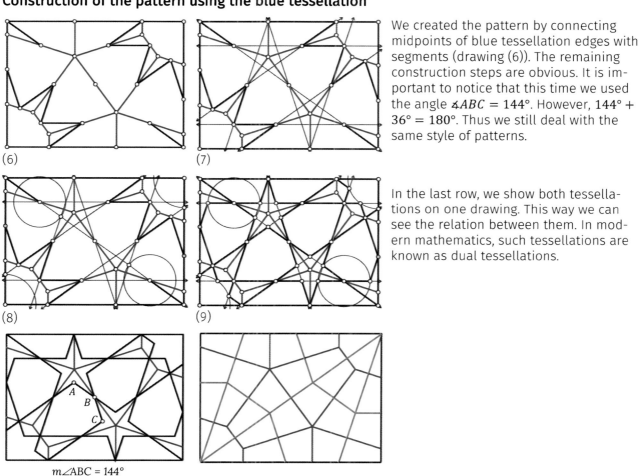

We created the pattern by connecting midpoints of blue tessellation edges with segments (drawing (6)). The remaining construction steps are obvious. It is important to notice that this time we used the angle $\measuredangle ABC = 144°$. However, $144° + 36° = 180°$. Thus we still deal with the same style of patterns.

In the last row, we show both tessellations on one drawing. This way we can see the relation between them. In modern mathematics, such tessellations are known as dual tessellations.

$m\angle ABC = 144°$

These two tessellations, the red one and the blue one, are not the only tessellations, which can be used to design our pattern. A few more tessellations can be useful for the pattern from project 3.3 and can be a source of ideas for many other decagonal patterns. On the next page, I compare all four of them.

Five tessellations for design 3.3

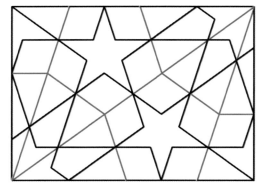

The red tessellation

As we remember, we created the pattern on this tessellation by drawing lines connecting opposite midpoints on the edges of pentagons. Patterns in the remaining figures (decagons and trapeziums) are a natural consequence of this choice.

This is the most popular and the most useful tessellation. We have here three polygons – a regular pentagon, a golden trapezium, and a regular decagon. The regular decagon can also be treated as a union of 10 golden triangles.

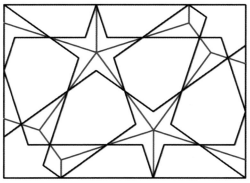

The blue tessellation

We created this tessellation on the previous page. For the pattern from project 3.3, its construction was very easy. However, for a more complex pattern, the construction of analogical tessellation can be a difficult task.

The pattern was created by connecting centers of edges of trapeziums. Note, the angles between the edges of tessellation and the pattern lines are not the same as in Devon Madrasah's style.

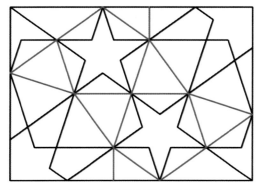

The green tessellation

This tessellation was derived from one of the Iranian books on geometric pattern design. It uses regular pentagons, regular decagons, and two types of triangles. The larger triangle we can fill with the pattern, but the smaller one we leave empty (no one ever said that all tessellation tiles should be filled with pattern).

The construction of this tessellation is simple. We use the vertices of polygons to create the pattern. This is the same approach as in the Sikandra style.

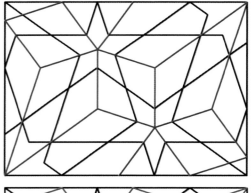

The purple tessellation

This tessellation was derived from a drawing in an old paper (1947) by Russian scientist Balkanov. Creating this tessellation is not a difficult task. The pattern was created by drawing lines through the midpoints of long edges of kites and parallel to the opposite long edges of kits.

NOTE: The white triangles in top-left and bottom-right corners are difficult to fill with the pattern following all gereh rules. Thus we had to break one of the rules of gereh design.

The violet tessellation

This tessellation was invented a few years ago by a French mathematician and pattern designer, J.M. Castera.

In this tessellation, we have two rhombi, one with angles 72° and 108° (fat rhombus) and another one with 36° and 144° (slim rhombus).

Chapter Summary

Let us summarize the information that we learned in this chapter.

TESSELLATIONS AND PATTERNS: The most important point is that from one tessellation, we can construct many different patterns representing different types or styles - depending on how we want to look at patterns. Here we see how true the saying is – every pattern is a decoration of the geometry behind it. Thus to create a pattern using the gereh technique, or method, we should prepare the right contour, a good tessellation covering this contour, and then add a pattern to this tessellation. As we will see later, some contours are good for some types of patterns but completely useless for other types. For example, it is difficult to create a decagonal pattern in a square contour.

From project 3.13, we may conclude that in many cases, one pattern can be created using a few different tessellations. In the red and blue tessellations, we used two different angles, 144 and 36 degrees, but these two angles are connected with the equation: 144° + 36° = 180°. In the same way, we can create the design from project 3.4. This time we will have angles 72 and 108 degrees and, of course, 108° + 72° = 180°.

The red and blue tessellations from project 3.13 in modern geometry are called dual tessellations. Each of them has its specific features and as we have noticed, constructing the blue tessellation was a more difficult task than the red one. In general, we will need to construct only one of these tessellations.

GEOMETRY: It is worth looking at the geometry of decagonal tessellations. The next drawing shows polygons that occurred in our tessellation. This is a very basic set, and it is convenient for many decagonal patterns. Later we will see other polygons that occur in decagonal tessellations.

> **MEMO**
> The same tessellation can be used to produce many different patterns.

> **MEMO**
> A few different tessellations can be used to create the same pattern.

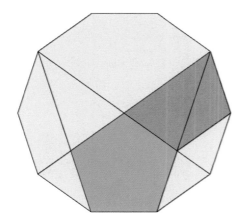

The drawing shows the relation between the three polygons used in the red tessellation for this chapter: decagon, pentagon, and trapezium.

One can easily calculate that these three figures have edges with two lengths only. If L is half the diagonal of the decagon and S is its side, then the pentagon has sides equal to S, and the trapezium has three sides equal to S and one equal to L. The long triangle that is 1/10 of the decagon area has two sides L and one S.

STYLES AND TYPES: In terms of different styles, we are still able to add a lot. In terms of types – many types in Tony Lee's manuscript are simply the same types as those mentioned in this chapter but using different tessellations. We will see this in a few later projects.

This discussion shows how difficult it can be any attempt at classifying geometric patterns. There is always more – we can find patterns mixing two or more different styles in one design. There are many such patterns in mosques from the Ottoman Empire. We already had such a pattern in this chapter. In project 3.7, we mixed the Tan Sahid style with the Persian style.

There is also another possibility. Depending on how we look at a pattern, we may classify it in one or another style. Finally, patterns may have several modifications or special decorations that make it hard to assign them to one or another style or type.

In the next chapter, we will see how we can customize our tessellation to produce more decagonal patterns.

4. Patterns on decagonal grids

It often happens that some well-known patterns are presented in a more elaborate form that gives them a new outlook. In this chapter, we will discuss a few such examples. Thus, the contour, as well as the tessellation, will still be the same. In some cases, patterns will be the same as in the previous chapter, but something extra will happen to them.

➢ Project 4.1 – Interlace patterns

The interlace form of patterns is very popular in marble carving. It adds a third dimension to the pattern – depth. This project will show how a regular gereh can be transformed into an interlaced form.

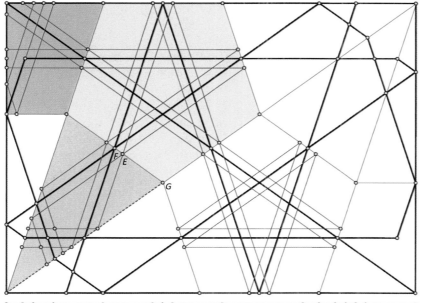

Fig. 4.1a. Here we start with the Kukeldash Madrasah pattern, but we could use any other pattern created in the previous chapter, excluding mashrabiya and Sikandra style patterns. In general, any pattern where we have an intersection of no more than two lines is good for interlace form.

We divide each edge of a pentagon into 12 parts. Here length FE equals 1/6 of the whole length FG (half of the pentagon edge). By joining the new points with segments, we will get a network of thin lines as here. Note: instead of 1/6, we can take 1/4 or something else that will be easy to construct.

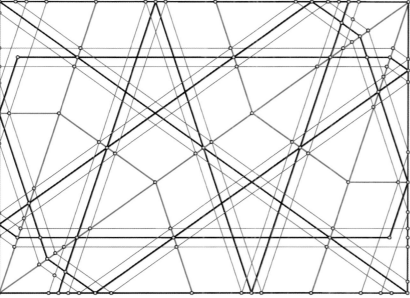

Fig. 4.1b. Here we show how the complete network will look.

The thick black lines are the lines of the original pattern, and in a moment, we will have to wipe them out. We do not need them anymore.

I have marked some key points that will help us to control our later work. But we do not need them while drawing this design on paper.

The thin blue lines are guidelines only, and we should draw them using a very soft pencil, so later we can easily rub them out.

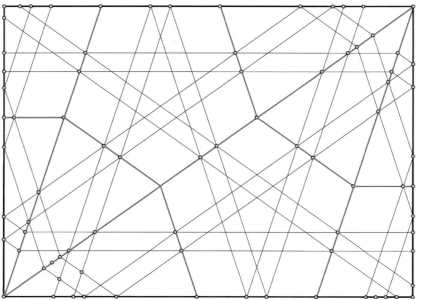

Fig. 4.1c. Here the supporting network of lines is shown. From this image, we can easily conclude that dividing the pentagon's edge into 12 parts gives us reasonably wide bands. If we divide it into 8 equal parts, then these bands will be even wider. In fact, they can be much wider. The only thing that we should care about are the small elements between the bands. They should not disappear completely.

If the bands are narrower, then the design may look a bit strange. This is really a matter of your taste.

Fig. 4.1d. Now, important – we take four copies of the template from the previous step. Remember, for the future – true interlace designs cannot be reflected. They never have any mirrors.

Now, start from the center and begin drawing in ink the thick black lines shown here. The bands of the new pattern should go up and down alternately.

A good option could be to develop the interlace pattern separately for each tessellation tile. This way, we may have smaller areas to control and fewer mistakes.

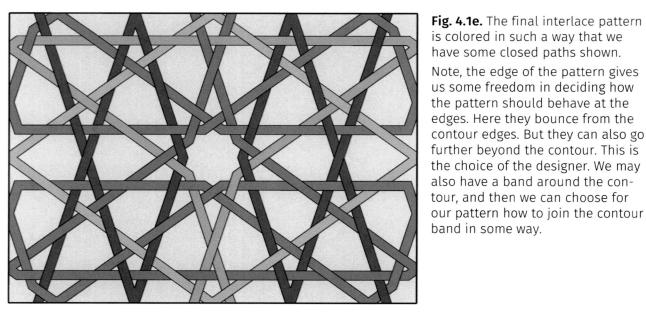

Fig. 4.1e. The final interlace pattern is colored in such a way that we have some closed paths shown.

Note, the edge of the pattern gives us some freedom in deciding how the pattern should behave at the edges. Here they bounce from the contour edges. But they can also go further beyond the contour. This is the choice of the designer. We may also have a band around the contour, and then we can choose for our pattern how to join the contour band in some way.

➢ Project 4.2 – Interlace patterns piece by piece

Drawing a large interlace pattern from four copies of a tessellation is a very tempting task. However, it often leads to many errors. Thus we may split this task into separate subtasks – each for a separate tile. Below, we show how to make such a design.

For this project, we will choose the pattern from project 3.5, the Tan Sahid Mosque style.

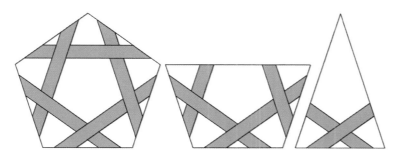

Fig. 4.2a. Each of the tiles of our tessellation can be covered with an interlace pattern separately. This way, we may avoid mistakes and simplify our work. Here we see how each tile should be covered with the pattern. Note – the bands in the pentagon and other polygons must rotate in the same direction.

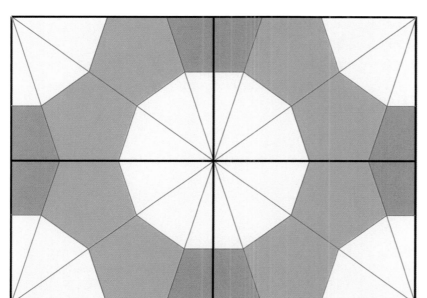

Fig. 4.2b. We start with 4 copies of our tessellation from the previous chapter.

Then we will fill each tessellation tile with a pattern exactly as it was shown above. In this project, the width of the bands on the pentagon edge is 1/4 of the length of the edge. It can still be slightly wider.

Fig. 4.2c. Here is a pattern created from a tessellation based on 4 copies of the original tessellation. Note how the interlace works near the common edge of two contours. The left and right sides are not mirror reflections of each other.

The same situation occurs between the top and bottom parts of the contour edge. This drawing demonstrates why interlace patterns do not have mirror lines.

Now, with a new template like the one here, we can produce a larger pattern. Again, no reflections. Translations only.

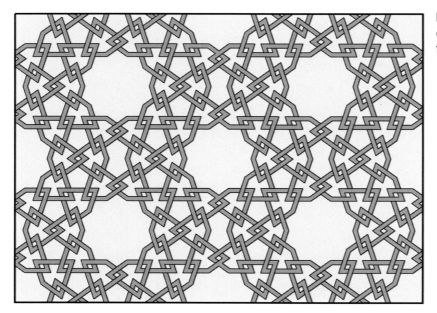

Fig. 4.2d. A large pattern based on 4x4 copies of the template from figure 4.2c.

Interlace patterns may have many forms. Its bands can be wider or narrower, have additional borders, or some extra decorations. Brian Wichmann described some of them in his book; see Wichmann (2018), chapter 11. We will do one more example with a decoration that is known as banding. In the next project, we will follow the example from the Gowhar Shad mosque in Mashhad.

➢ Project 4.3 – Banding Mashhad style

For this project, we will use the gereh from project 3.4 – the Persian style. This is not exactly the same as the original gereh from Mashhad. It is much simpler and less demanding in terms of labor amount.

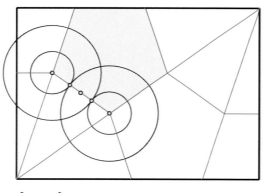

Fig. 4.3a. We start with the same tessellation we have been using up until now. In one of the pentagons, we divide the edge into two equal parts, and then each part, we divide again using the golden section. In the next step, we will divide the other edges of the pentagon in the same way. This can be done using circles passing through these points and with centers in the pentagon's vertices.

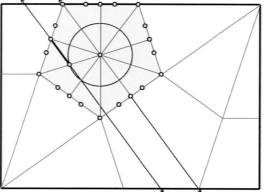

Fig. 4.3b. Here I show how to start the construction of a star from project 3.4. This is just a quick reminder only. You should know this part already. The two black lines are parallel. The circle will be used to create the star inside a pentagon.

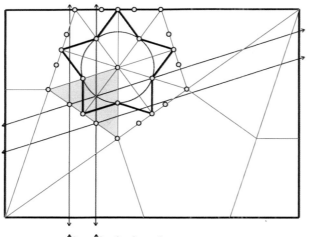

Fig. 4.3c. The star from project 3.4 is ready. We use these additional points obtained by the golden section to draw lines parallel to the star's edges. Here we do this only for 1/5 of the pentagon. If we draw all lines parallel to all the star edges, we will have too many lines and a terrible mess.

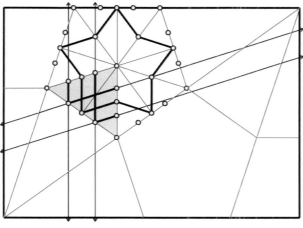

Fig. 4.3d. We use parallel lines to create 1/5 of the pattern inside the pentagon. We should repeat the same in the other parts of the pentagon. If you want to simplify your work, you can copy this pattern to the other parts of this pentagon and to the other one.

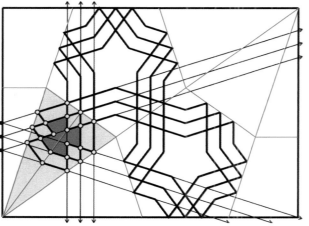

Fig. 4.3e. Here both pentagons are filled with the pattern. Now we can concentrate on one of the triangles from the decagon.

The dark shape inside the triangle is only to help us see the pattern's structure. If you do not like it, you can remove it.

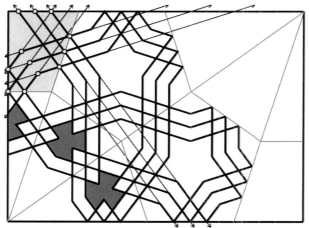

Fig. 4.3f. Here I show how the half trapezium in the corner can be filled with the pattern.

Now you can fill the remaining parts of the tessellation with the pattern. Think about coloring it, etc.

Note – this is not an interlace pattern. Thus, we can use mirror reflections about the edges of the template to make a larger pattern.

4. Patterns on decagonal grids | 49

Fig. 4.3g. This image shows one of the possible patterns made using this approach. Here we changed the pattern slightly inside the trapezium. Can you find out what the change was and how it was made?

If you wish to experiment with this design, the grid used in this project is enclosed below. Its size is 12x8.82 cm, which is enough for a quarter of an A4 page with a decent margin. Thus you can make a full A4 page design using various combinations of lines, fills, and coloring.

There are numerous patterns made on decagonal grids. Some of the designs made on such grids can be very attractive. Several such patterns we can find in Iran, especially in Yazd and Isfahan. Very simple patterns made on similar grids can be found in Uzbekistan and even in India. Let us try some of them.

➢ Project 4.4 – Decagonal grid pattern Tony Lee style

This example does not come from any real location in the world. It was invented by Tony Lee, and it can be considered a preparation for more complex designs. This is something that we need now before we jump into Yazd style patterns. For this pattern, we need four copies of our tessellation. If you use one copy only, then you will have to use mirror reflections in your pattern. Taking four copies, as for the interlace patterns, will allow us to use translations to make a larger pattern.

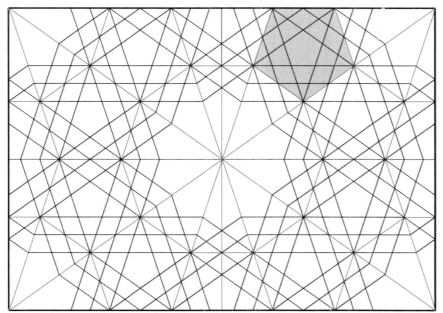

Fig. 4.4a. The basic decagonal grid

We start with four copies of the tessellation. Then we find the center for each pentagon. Finally, we draw lines through the midpoints of the edges of the pentagons and their vertices.

Note – we have here bunches of three lines rotating around the central point of this rectangle. These lines change their directions on the red lines of tessellation. Thus we are still following gereh rules.

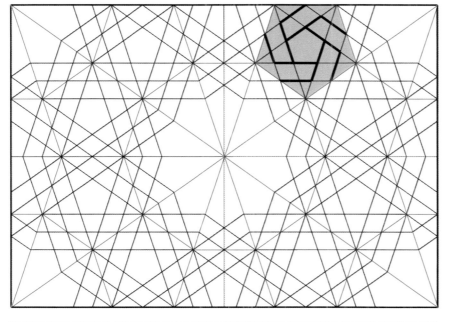

Fig. 4.4b. Here is shown how we can create a kind of swirl inside one of the pentagons. Now, we have to repeat exactly the same drawing in the remaining pentagons.

Fig. 4.4c. After filling all the pentagons with an identical pattern, we also have to fill the six trapeziums.

The pattern for the trapezium can be obtained from the pattern in the pentagon. Parts of the pattern marked with thin lines should be removed.

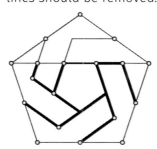

4. Patterns on decagonal grids | 51

Fig. 4.4d. Drawing the pattern inside the central decagon is quite an easy task. We have to follow the external grid lines in the decagon. The gridlines that are closer to the center we may leave free or use them in some way.

Fig. 4.4e. Finished pattern by Tony Lee. It can be treated as a final pattern, or we can use it as a template for larger projects.

Now we have a quite clear image, and you can think about how to paint it nicely. Usually, such mosaics are made out of pieces of ceramic. Thus we should avoid putting the same colors next to each other or separate them by strong bands. The pieces shown here can be quite difficult to cut them from ceramic tiles. For this reason, each of them was split into smaller chunks that are easier to cut.

Fig. 4.4f. Here is a larger pattern created from the template that we created in the previous figure. We used 4 copies of the template and translations along its sides to have continuity of lines on the border between each two copies.

If you want to experiment with this design, you still have many options. For example, splitting the pentagon into shapes can be done in several ways. We can also consider the swirl in left and right directions alternatively.

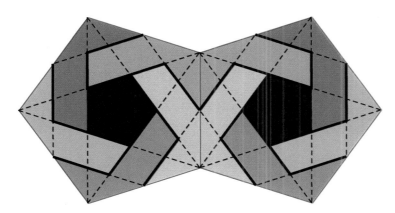

Fig. 4.4g. Another version of connecting patterns from two pentagons. This time the left side is a mirror reflection of the right side. This way, we will produce several elements that will have one mirror symmetry. In the previous arrangement, we had only rotations and no mirrors at all.

➤ Project 4.5 – The Yazd style decagonal grid

A group of patterns mostly from Iran uses a similar approach to the one used in the previous project, but the pentagon's center contains a star. Each of them uses a very specific decagonal grid. This project aims to investigate this type of grids and make them ready for the next project.

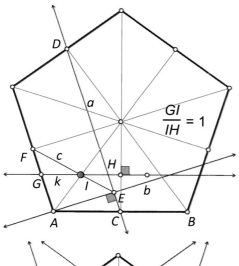

Fig. 4.5a. Yazd slice of a pentagon

This construction will be useful for creating Yazd decagonal grids.

We start from a regular pentagon and all its symmetry lines (red segments).

Draw line 'a' connecting the midpoints, C and D, of the pentagon's two opposite edges.

From the bottom left corner, draw line 'b' perpendicular to 'a'.

Draw a segment connecting points E and F. The intersection of this segment with the pentagon's symmetry line was marked as a large red point I. Draw a horizontal line through this point (line 'k').

It can be proved that |GI|=|IH|.

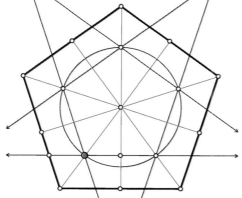

Fig. 4.5b. Yazd slice of a pentagon (cont.)

Remove all construction elements from the previous step. Leave only the red point and horizontal line passing through it. Draw a circle with its center in the center of the pentagon and passing through the red point. Draw lines passing through the points of intersection of the circle with diagonals of the pentagon.

This way, we will obtain another pentagon inside the original one. We will call it the Yazd slice of a pentagon.

Now we are ready to create the Yazd decagonal grid. In fact, it is possible to create a few similar grids.

We can start with the same tessellation that we have used in all previous projects. As you remember, this tessellation has only three different tiles – a regular pentagon, a trapezium, and a quarter-decagon. The decagon can be split into long triangles. This way, we can deal with smaller tiles.

To simplify the design of the decagonal grid, we will show how each tile should be transformed. Then we will assemble all the pieces together.

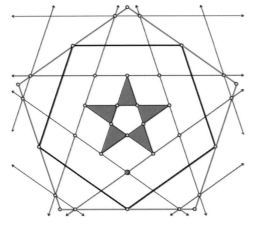

Fig. 4.5c. The grid in a regular pentagon

Take a regular pentagon (red segments); connect the midpoints of edges to get a smaller pentagon (thicker lines). Apply the Yazd slice to this pentagon and draw a perfect star inside it. Here I filled the star with some color to emphasize its position. You do not need to follow my coloring scheme at all.

Note – the purpose of the Yazd slice was to get all slices of the same width. This was important.

Fig. 4.5d.e. (below)

The left drawing shows what happens if we want to apply the same grid to the trapezium. The star in the middle will be cut in the wrong place. But, there is no way to do it differently. Thus here, we have to cheat by making the top petal of the star shorter.

The right drawing shows the grid inside the long triangle. It is a natural extension of the grid from the pentagon.

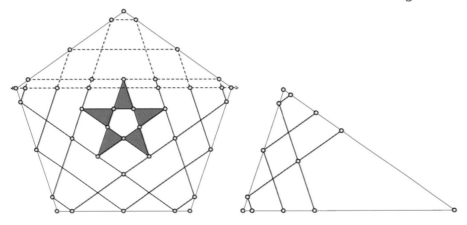

Now, we have created all pieces, and we can fill the tessellation with them. At the end of this book, in the appendix, a large Yazd grid is enclosed. It is around 16 cm wide (depending on how this text will be printed). Thus you can use it for A4 designs without having to construct it again and again. If you want it larger, then you will have to construct it at the right size. Yazd style patterns are very attractive. Thus it is really worth making a large design.

➤ Project 4.6 – The Yazd style swirl with stars

The next photo shows one of the Yazd patterns. There are a few more of them, and one can create his/her own patterns in the same style. It is important to notice that a tessellation for the Yazd style pattern should contain as many regular pentagons as possible. The swirl inside the pentagon makes this type of design very attractive. It is also worth noting that the adjacent pentagons' pattern can have the same orientation or opposite (see fig. 4.4g). In the first case, we will not have mirror symmetries in our pattern. In the second case, we will have several shapes with one mirror symmetry line.

Fig. 4.6. A pattern from the Jame Mosque, from Yazd in Iran, has a ring of pentagonal stars around the large central star. The pattern is made out of ceramic pieces with a dark blue line underlying their shapes. Note how the pattern is built between the two adjacent stars. Note also, the central star on top and in the bottom is somehow flattened. Do you remember why?

Now, we can start developing the pattern from Yazd. Again we will discuss the pattern using separate tiles of our tessellation. This way will be much simpler, and definitely, our drawings will be less crowded.

Fig. 4.6b.c.d. The three tessellation tiles are filled with patterns. The thick lines are emphasized lines of the pattern; the thin lines should be erased after finishing the whole design.

The pattern for the trapezium can have one symmetry line – a vertical mirror, but then we will have to adjust its pattern also. Note the photo from Yazd shows it exactly as it is here.

The long triangle has a minimal pattern, exactly as in the photo. But it can be more elaborated by adding internal rings of extra bands.

4. Patterns on decagonal grids | 55

Fig. 4.6e. Pattern from Yazd with all pentagonal and trapezium tiles oriented in the same direction.

Fig. 4.6f. Pattern from Yazd with all pentagonal and trapezium tiles oriented in alternate directions. Here we can easily distinguish elements that have one mirror symmetry line.

An extra ring was added inside the big star. It occupies extra space. By adding a few more such rings, we can fill this star completely.

➢ Project 4.7 – The Yazd style swirl with loops

Fig. 4.7a. This is a 'do it yourself' project. The pattern shown here is a slight modification of the Yazd pattern from the previous project.

Your goal will be to trace the changes and reconstruct this pattern completely on your own.

Coloring this pattern can be done in a very simple way or in a more structural. You have probably noticed that some elements of this design can be treated as a pseudo-interlace pattern. Thus you can find here long tracks interlaced with other tracks. While using the same color for each tile of a track, we can emphasize it.

➤ Project 4.8 – The Yazd style with some extensions

Yazd patterns are very attractive due to the pentagonal swirl with a star in the middle. As we know, the swirl comes as a filling of a regular pentagon. Thus, if we can produce a tessellation with many pentagons, we will get a more complex Yazd style pattern. Here is one of such designs.

Fig. 4.8. Yazd style pattern from Shah Mosque in Isfahan

This pattern has two parts. The top one is the same as we did in our projects. The one to the bottom of the central star is definitely different, although it uses the same approach – pentagons with the same swirl. In this example, we will learn how we can construct new patterns Yazd style.

NOTE – for the sake of convenience, we will construct this pattern rotated 90 degrees. In this project, we will reconstruct only the bottom part of this pattern.

Before we dive into the pattern's design, we should construct a new contour and then make a tessellation for it.

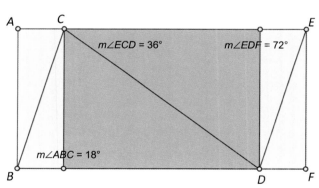

Fig. 4.8a. Construction of the contour (rotated 90°)

This construction shows how we can create the contour for this project. We start from segment AB. Then we construct three rectangles. The left one and the right one are identical. We created them using 18=90/5 degrees angle. The middle rectangle is the same one that we used on the previous projects.

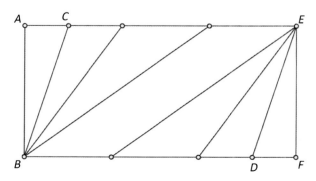

Fig. 4.8b. To produce the tessellation, we divide the opposite angles into 5 equal parts. The longest section lines are not used in this construction. For this reason, we omitted them.

Note, the points on the long edges of the contour give us all that we need to create similar decagons and pentagons to those in the previous projects' tessellation.

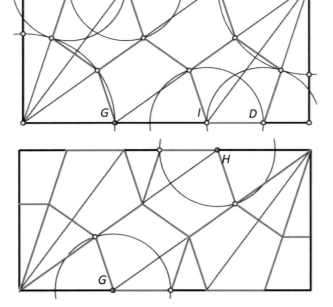

Fig. 4.8c. This drawing shows what we get from the points on edge. Points G and H are anchors for creating quarters of decagons in the opposite corners of the contour. Then circles in Points C, J, I, and D are enough to develop the two pentagons.

Fig. 4.8d. Here we show the last step for finishing the tessellation. We have here four regular pentagons, half trapeziums, and quarters-decagons. Moreover, we have here two narrow triangles that are usually avoided in decagonal pattern development. These shapes are new for us, and we will have to deal with them in a moment.

In the next chapter, we will come back to how we can create contours and tessellations. Now let us see how we can fill this tessellation with a Yazd style pattern. Most of the work is done already. We have our new tessellation. We know how we can fill pentagons and long triangles with the pattern.

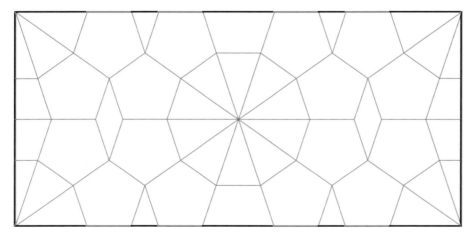

Fig. 4.8e. Here we have a large tessellation allowing us to make a design without symmetry lines.

We took four copies of the tessellation that we created in the previous step.

Fig. 4.8f.g.h. Filing this tessellation with the pattern that we used in the previous project gives us this design. Our task will be to fill the narrow triangles that have now changed to narrow rhombi.

The next image shows how to construct the pattern in the narrow rhombi. We insert exactly the same pattern in all narrow triangles and rhombi.

Fig. 4.8.i. Pattern based on the bottom part of the design from Shah Mosque in Isfahan

The final pattern for this design may look like this. A few other variants can be done if we use a different approach to the local symmetries, e.g., use mirror symmetries. Even small modifications of the pattern in the pentagon may change the whole design significantly.

In the next chapter, you will find a few decagonal tessellations with regular pentagons. You can try to use some of them to experiment with Yazd style patterns. Meanwhile, let us try to make a very unusual pattern.

The detailed construction of this tessellation is given in the next chapter (see tessellation TG1). The pattern can be developed the same way as we have shown, or you can use tiles oriented in alternate directions.

➢ Project 4.9 – A monster pattern Yazd style

Below we have two figures. The left one shows a complex tessellation with many regular pentagons. The right figure shows a Yazd style pattern created using this particular tessellation. Your task will be to design it (see tessellation TG1 in the next chapter), and copy four instances of this tessellation on a large piece of cardboard and then develop the pattern shown here.

The Yazd style patterns were also developed with hexagonal and other symmetries. Although none of them was as complex as those using decagonal tessellations.

Geometric pattern from Hakim Mosque in Isfahan

This pattern combines dodecagonal and octagonal local symmetries, and at the same time, it uses the Yazd style approach. Pentagons in tessellation for this design are not perfect but still good enough to obtain a nice regular view.

5. Contours and tessellations

Geometric patterns were not created for the sky or the universe. They were created for limited spaces with reasonably fixed size and proportions. Imagine a kundekari master creating patterns for the doors of a mosque. The size of the doors was fixed already. He could not change it. The frame around the decoration on the doors had a fixed width proportional to the other elements. Thus, he had to produce a pattern that will fit reasonably well into the designated area. He had some subtle ways to adjust the space for the main pattern but within limits. Thus, he needed to precisely calculate the size of his pattern as well as its all elements.

Fig. 5.1. Doors to the mausoleum of Sultan Ahmed in Istanbul

This photo shows the relations between the main components of kundekari doors. We have three parts – the central pattern, the bottom pattern that is usually derived from the central pattern, and the top part that often contains some extra geometric pattern or calligraphy. The central pattern and the bottom one have fixed proportions. The top panel with calligraphy allows the designer to adjust the space for the two other parts. But this part should also retain reasonable proportions. It cannot be very flat or very high.

Frames around every three parts of doors have the same width all over.

Note also – each element of such pattern is a separate piece made in the wood often decorated with mother of pearl. Finally, we have some narrow wooden frames around each piece.

As one can now notice, creating a geometric pattern for kundekari doors was a strict engineering work. Usually, the designer started from the pattern's width and then created a contour using divisions of the right angle. Then he constructed a pattern to fit into his contour. This was not the end of his work. Based on the pattern's proportions, he had to calculate each shape's size that would be used in the kundekari doors and then the sizes of frames around each shape. Thus, the proportions of the patterns are the most important thing to start with. In modern terms, we call this top-down design. This kind of design is used by contemporary architects and engineers.

Decagonal contours

In this section, we will present only a few of the most important contours for decagonal patterns. We already have been using two of them. There is also the arithmetic of contours. We can add them in many ways as well as subtract one contour from another. This last operation is not often used. Let us start with a useful notation.

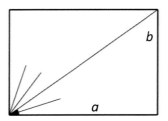

$|a|/|b| \approx 1.3763 \approx 1.38$
Here $|a|$ and $|b|$ mean lengths of segments a and b.

Fig. 5.2. Contour C(2/5).

Here we have the simplest and the most popular contour for decagonal patterns. We used it in all the projects in previous chapters. Note how the contour was created. We start from the base segment and two lines perpendicular to the base at its endpoints. Then we divide one of the angles into 5 equal parts, and we draw a horizontal line where the second section line intersects with the right perpendicular line. This creates a contour that we will represent by the formula C(2/5).

Fig. 5.3 Contour C(1/5)

This is a very flat contour often used to expand another contour up or down. The proportions of this contour are $|a|/|b| = 3.0776 \approx 3.08$.

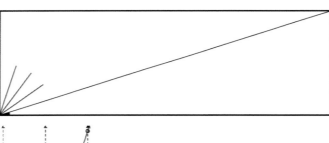

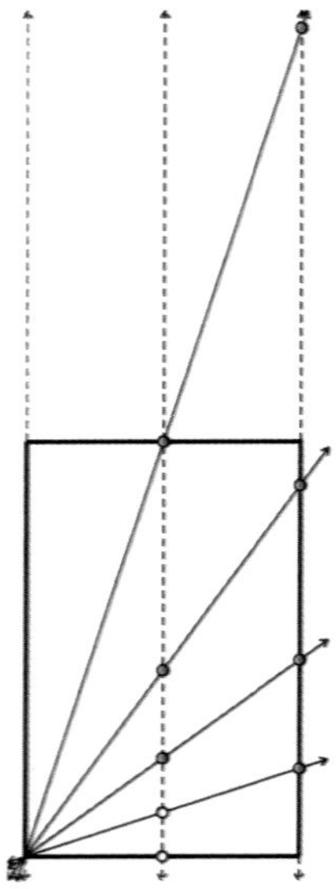

Fig. 5.3a Ways of looking for decagonal contours

Any of the red points shown here and many others constructed this way can be used to create a decagonal contour. The only problem is if we find a way to create a tessellation covering contour created this way.

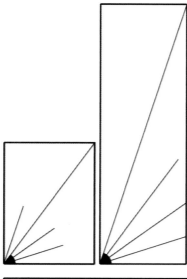

Fig. 5.4 Contours C(3/5) and C(4/5)

Contour C(3/5) is the same as C(2/5), rotated 90 degrees around one of its vertices. It is often used for vertical patterns. Contour C(4/5) is the same as C(1/5), rotated 90 degrees around one of its vertices. Many kundekari doors in Turkey and Iran use this particular contour for patterns on doors.

Here $|a|/|b| \approx 0.7265 \approx 0.73$ for the left contour and $|a|/|b| \approx 0.3249 \approx 0.32$ for the right one.

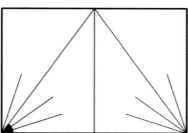

Fig. 5.5 Contour C(3/5)+C(3/5)

As we mentioned, we can add contours. Here we added two instances of C(3/5). A contour like this is often used for some wide patterns that do not fit well into C(2/5). Here $|a|/|b| \approx 1.45$

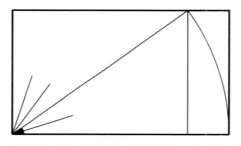

Fig. 5.5. Contour C(2/5)+C(4/5)

This nice contour can be created as C(2/5)+C(4/5). Its proportion is

$$|a|/|b| \approx 1.70$$

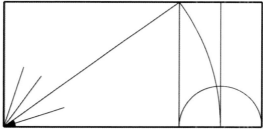

Fig. 5.6. Contour C(2/5)+C(4/5)+C(4/5)

Here we see how we can further expand contour from the previous figure. Of course, we could still go further and further. Here $|a|/|b| \approx 2.03$.

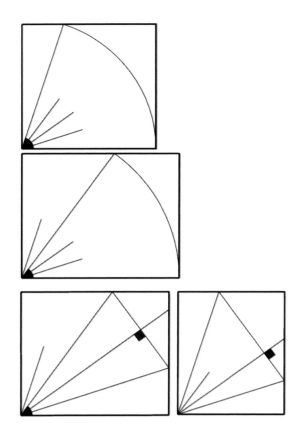

Fig. 5.7a.b. Contours C*(1.05) and C*(1.24)

The left contour is almost square and can be used for decagonal patterns covering an almost square area. Surprisingly, many patterns are using this contour.

The right contour is similar to C(2/5). According to Rempel, it was the most frequently used contour in XVI-XVIIc.

Fig. 5.8a.b. Contour C*(1.1756) and C*(0.85)

The left contour is rarely used but convenient for some unusual patterns. Its proportions are

$$|a|/|b| = 1.1756 \approx 1.18$$

The right contour is often used in kundekari art.

$$|a|/|b| = 0.85$$

For the last few contours, we used the simplified notation C*(number), where 'number' is a decimal number expressing the exact proportion of the contour. The reason is that these contours only partially depend on divisions of the right angle into 5 equal parts.

There are numerous contours and many ways of combining them together. We will see more of them in a few projects in the next chapter of this book.

Selected basic decagonal tessellations

Tessellations used for gereh patterns can be very simple, but it often happens that a simple tessellation with a few tiles can be a starting point to a monster tessellation with hundreds of tiles. In the next pages of this chapter, we will show some of the simplest tessellations. We divided them into four groups, but you can organize them in any other way.

Group A. The most frequently used simple tessellations

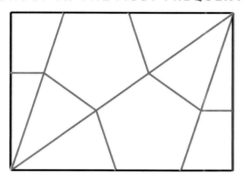

TA1 The most popular tessellation for decagonal patterns. A very large number of patterns use this tessellation in some way. I wrote here 'in some way,' and the ways of using it can range from simple ones, like those in chapter 3, to very sophisticated ones.

From chapters 3 and 4, we also remember that we could create several different style patterns on this particular tessellation. Thus, it is one of the most universal tessellations.

In the Topkapi scroll, we can find a few applications of this tessellation.

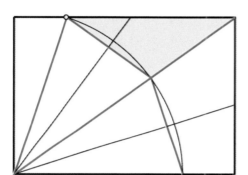

TA2 Here is the simplest tessellation for contour C(2/5). It contains a quarter of a regular decagon, half of a pentagon, and half of a rhombus. We already know how to deal with decagons and pentagons. This rhombus is a new tile for us. It will work excellent with some styles and not well with others. All of its sides are equal.

This simple tessellation will also be very useful in complex designs.

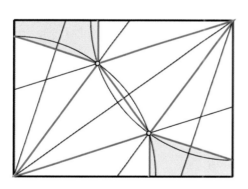

TA3 This is another popular tessellation. Many patterns all over the world use it. We can also find it in the Topkapi scroll. This is one of the most useful decagonal tessellations. We can modify it in many ways and produce patterns with large complexity. It can be used for a few different styles. According to Rempel (1961), it was one of XII's favorite tessellations in Central Asia.

In this tessellation, we have half of another new polygon – a tall trapezium. It is a very useful polygon. Note, its three sides have equal length.

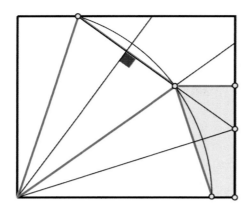

TA4 This tessellation uses the same polygon as the previous one – the tall trapezium. The contour was created by drawing a line from the end of the first section line of the 90 degrees angle perpendicular to the third from the bottom section line. Again, according to Rempel, this tessellation was very popular in the XIII century in Central Asia. We can find it also frequently used in Turkey, for example, in the Beyhekim Mosque (1270), the Karatay Madrasah in Konya, the Sultan Hani in Aksaray, and many other places.

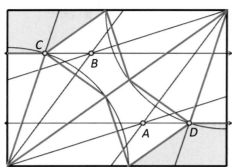

TA5 This tessellation I found in a pattern from Termez, XII century. We see here the tall trapezium that we discovered in the previous example. We also have here the quarters of a new polygon – an elongated hexagon (light background). We will see it better in the next tessellation.

The labeled points show the order of how this tessellation was constructed. It uses contour C(2/5). Lines passing through points A and B are parallel to the bottom edge of the contour. Their intersections with section lines of the 90 degrees angles (points C and D) are used to create quarters of the two decagons.

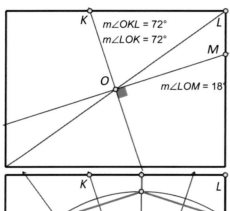

The construction of this tessellation needs a bit more work. We start with contour C(2/5). Through the midpoint of the diagonal, we draw two lines. One passing through points O and K has a 72 degrees angle with the diagonal, and the other one passing through points O and M is perpendicular to OK. Points K and M are the key points for the rest of this construction.

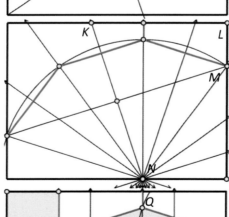

We divide the angle around point N into 10 equal parts. In fact, we do not need all these section lines. We need only every other one of them.

The circle with radius NM and center in N gives us the intersection points with the section lines marked here. This is enough to draw the three red lines. This is the beginning of our tessellation.

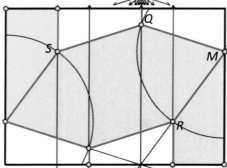

TA6 Two more circles with centers on the right and the left edge of the contour allow us to draw the remaining part of a long hexagon. This is the same one that we have seen in the previous tessellation.

Note: this hexagon has all edges equal and angles 72 and 144 degrees (144=2*72). In modern mathematics, polygons with all sides equal and two angles occurring in some regular order are called shields. Shields were frequently used in geometric patterns, but at this time, no one used this particular name.

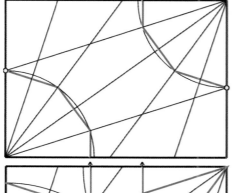

Here is a series of very useful tessellations. Some were taken from the Topkapi scroll.

STEP 1
We start again from contour C(2/5) and section lines dividing opposite corners into 5 parts. The points marked on the right and left edge can be used to create two quarters of decagons. From this moment, our construction can go in many ways producing different tessellations each time.

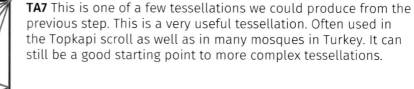

TA7 This is one of a few tessellations we could produce from the previous step. This is a very useful tessellation. Often used in the Topkapi scroll as well as in many mosques in Turkey. It can still be a good starting point to more complex tessellations.

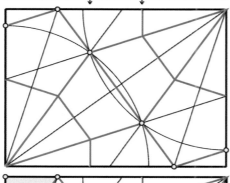

TA8 Here is another outcome from STEP 1. It is interesting to see four pentagons in one small tessellation like this. One of the problems while creating patterns with it, can be the small triangles in the left-top and right-bottom corners. But at least this can be a very useful tessellation for Yazd type of patterns. As you remember, we did not have problems with such a triangle there (a quarter of narrow rhombus).

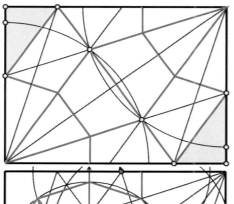

TA9 This tessellation is just a variant of the previous one; the only change was to cut 1/3 of the two pentagons and produce two trapeziums. But then we got the quarters of rhombi (shaded areas). These two rhombi are not the same as those that we constructed in the second example. Their side is L, while the other one could have side L or S depending on how we decide to construct the pattern.

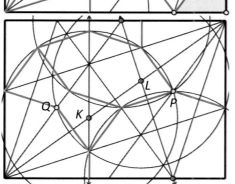

If we come back to step 1, we also can proceed differently. By drawing all the supplementary lines seen here, we use the red points as the centers of circles passing respectively through points P and Q.

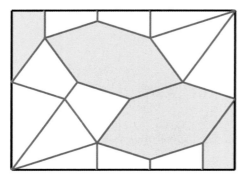

TA10 This tessellation we will obtain by finishing construction from the previous step. We can find this tessellation in the Topkapi scroll also. It is a very useful tessellation for many patterns. Note: here, the long hexagons can have sides equal to S as well as L. If the hexagon sides are L, then we have decagons with sides equal to L and sides of long triangles equal to X. Thus, we will get two monster decagons and very complex patterns.

Tessellations using long hexagons can often be created using a method that I call – tangent and overlapping decagons method. We will come back to them in a while.

This is the contour that we made already as C*(1.05). One can easily notice that it can also be created as a sum of two contours C(4/5)+C(3/5). Its shape is close to square. A few tessellations are using this contour.

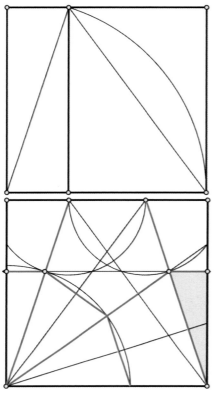

TA11 This drawing shows how the contour C*(1.05) can be used to construct a tessellation. Here we have one polygon that until now did not occur alone. This is a long narrow triangle. Here we have only half of it. Usually, this triangle is a result of the division of a decagon by its diagonals. Here we got it as a separate polygon. Such triangles alone or in groups are very useful in Kukeldash Madrasah style patterns. While developing patterns in Persian style, a single triangle like this makes design almost impossible. However, two such triangles together, side by side with a common vertex in the narrow end, are desirable in some styles. The tessellation shown here can be found in the Ankara Ethnographic Museum on kundekari doors exhibited there.

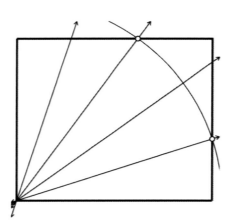

The construction of this contour is very simple – start with the point on the right edge, draw a circle passing through this point with its center in the opposite corner. Get the point of intersection with the third section line, and draw there the top edge of the contour.

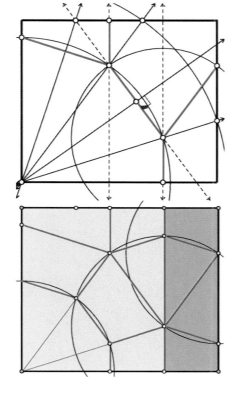

TA12 The tessellation constructed this way contains half and quarter of a long hexagon and a few other shapes introduced in previous examples. This tessellation was used in the Akbar's Mausoleum in Sikandra. The beauty of this tessellation is that it can be easily expanded to the right, forming a sequence of long hexagons as many as we need. This can be seen in the image TC2 (check section C). The number of long hexagons inserted there can be any. Depending on how many we add them, we may have the quarter of the new decagon on top or the bottom of the tessellation.

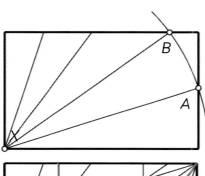

TA13 This is a variant of **TA12**. There are two interesting things in this example.

(1) It shows how one can transform an existing tessellation into another one more complex. This idea will be discussed in one of the last chapters of this book.

(2) The two shaded areas show us that we can add or subtract a tessellation to or from another. This tessellation can be considered as a union of the two tessellations shaded with different colors. The right one, darker, we will discuss later.

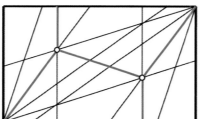
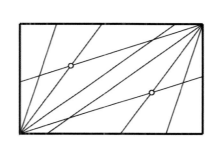
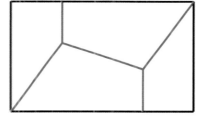

TA14 Tessellation with two pentagons and two hexagons

In this tessellation, the contour is unusual. We take point A and draw the circle c(X,XA). This way, we get the point B. Here goes the top edge of the contour.

The two points shown in the next drawing are used as endpoints of a common edge of two tangent pentagons.

This tessellation can be used to make a very large mosaic without a rosette, only pentagonal stars.

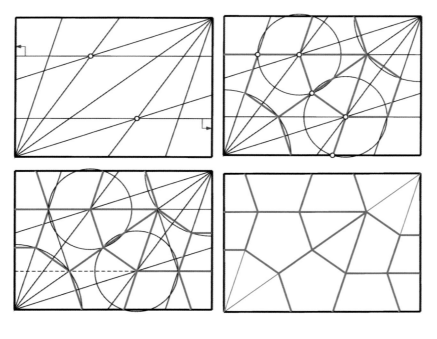

TA15 This interesting tessellation can be considered as a natural extension of the standard tessellation. Its creation starts from standard contour, and these two points shown in the first drawing.

The remaining stapes are similar to the construction of the standard tessellation.

It is a popular tessellation used for many patterns all over Iran, Middle East, and Turkey.

Group B. Tessellations with long contour

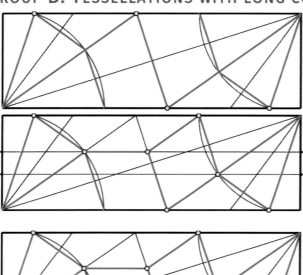

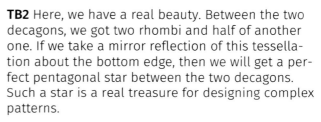

TB1 This tessellation was constructed on contour C(1/5). It is a very useful tessellation for long narrow patterns – horizontal or vertical. It was also used to create more complex tessellations.

TB2 Here, we have a real beauty. Between the two decagons, we got two rhombi and half of another one. If we take a mirror reflection of this tessellation about the bottom edge, then we will get a perfect pentagonal star between the two decagons. Such a star is a real treasure for designing complex patterns.

TB3 This is only a slight modification of the previous construction. But we got something completely different. There is no pentagonal star with rhombi, but we have two long hexagons and two long fillers.

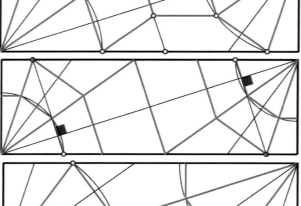

TB4 This is a specific tessellation combining two trapezium sizes – a regular and a large one with edges length X and L. It is very interesting to see how different the patterns in both types of trapeziums will be. We will have to come back later to this tessellation.

TB5 The same contour but a different choice of section lines to draw parts of decagons produces a very simple tessellation. Note – we have two halves of pentagons, but we do not have the quarters of a decagon. Why?

Group C. Tessellations using combined contours

There are many tessellations and patterns created on combined contours. Sometimes it is convenient to add two or more contours together and make a tessellation on such a new contour. In the next few examples, I will show how it can be done.

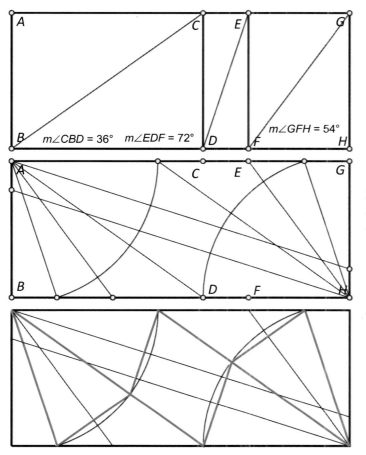

In this example, we have three different contours combined together. We start from segment AB and construct on it the contour C(2/5). Then we construct on its right side contour C(4/5) and finally C(3/5). Thus, we have here C(2/5)+C(4/5)+C(3/5).

In this step, we divide the two opposite angles into 5 equal parts. Using the first section line from the left and right, respectively, we draw the two arcs. This is all that we need to create a nice tessellation.

TC1 Here is a tessellation with two quarters of decagons and three rhombi. This tessellation comes from the Topkapi scroll. It can be further modified to produce a more complex tessellation.

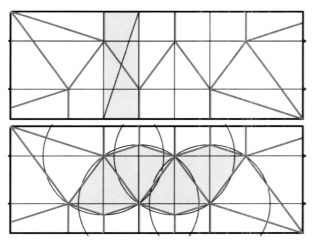

TC2 This is one of a few expandable tessellations. Note – we have here on the left and right side the same contour and tessellation that we have already introduced. Then, by adding multiple instances of the contour C(4/5) between them, we may produce a pattern that is as long as we need it. A similar tessellation was used in Sultanhani in Aksaray.

TC3 Here, we have a slight modification of the previous tessellation. It is also expandable to the right and left. We obtained here a new polygon – a wide kite. This kite was often used in Seljuk and Ottoman decagonal patterns. It has two sides L and two S. We can find a pattern using this particular tessellation in Istanbul, in Kılıç Ali Pasha mosque.

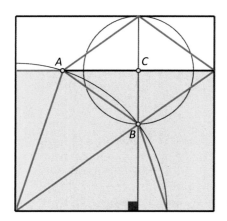

TC4 Here is another expandable tessellation. We start from the shaded contour C(2/5). Then we draw a circle passing through point A on the top edge to get a quarter of a decagon. Then, after drawing a line perpendicular to the base through point B, we get point C. The circle with center in C and radius CB allows us to extend contour upwards.

This particular contour is almost square. We have proportions |a|/|b|=0.95906 (where 'a' is the base and 'b' is the side). This makes it the closest approximation of a square between decagonal contours.

Note also – this design is easily expandable upwards. On top of it, we have the halves of a long hexagon and rhombus. By adding more hexagons and rhombi, we can go up as much as we want.

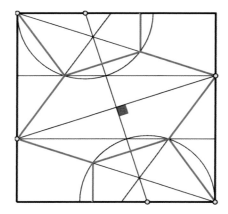

TC5 This nice tessellation was made using three stacked contours C(1/5). Its construction is quite obvious. It comes from a pattern in the Topkapi scroll. It was also used for a few patterns in Turkey. It can also be used as a starting point for a group of complex tessellations with many tiles.

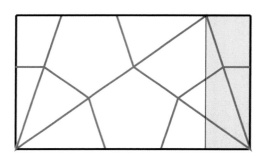

TC6 Here, we have a tessellation created on the combination of two contours. The left one is C(2/5), and the right one is C(4/5). A tessellation like this was often used in small geometric patterns on the bottom of doors and window shutters in Ottoman mosques and mausoleums. Two copies of this tessellation can be used to create a fairly complex and non-trivial pattern. Note – the single long triangle on top of this tessellation makes it useless in Persian style designs. But if we have two copies of this tessellation, with the top edge of it as a mirror line, we can produce an excellent Seljuk style design[1].

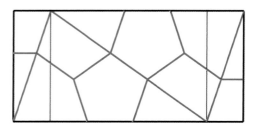

TC7 This tessellation is an extension of the standard tessellation that we used from the very beginning of this book. We can extend it further to the right or/and left by adding some extra narrow or wide contours. Its construction should be done from a vertical side of any of the rectangles shown here.

---

[1] Until this moment, we didn't talk about Seljuk, Samarkand, and Ottoman style designs. We will discuss them in another book.

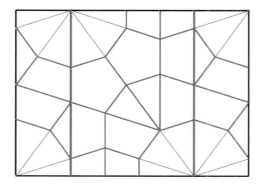

TC8 On the first look, this tessellation does not look very simple. However, if we draw the shown here thick vertical segments, we can easily notice that we have here the union of two tessellations that we already created. The central part was created as TA14. The narrow side parts are copies of the tessellation TB1. This way, we got a very interesting geometry for a pattern with two rosettes or large stars on the top and bottom edge.

GROUP D. TESSELLATIONS WITH LONG KITES

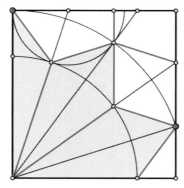

TD1 In this tessellation, the red points show how the contour was created. The other points show how the tessellation was constructed.

We have here a new polygon that we did not have before. It is a long kite with a 36 degrees angle in the narrow end and 72 degrees at the other end. This is a very rare polygon in geometric pattern design, but it is useful in Samarkand style designs and many Turkish designs. Patterns using this particular polygon are often made wrong. The tessellation shown here is one of the simplest tessellations with this long kite.

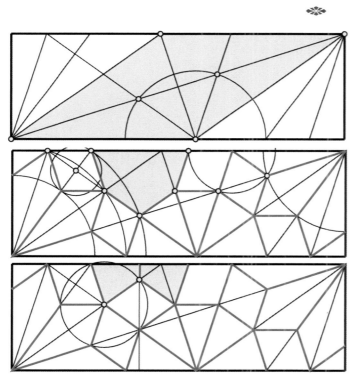

This will be another tessellation with long kites and a shield from the Topkapi scroll that we have seen in chapter 1. We start with long contour C(1/5). Section lines of two opposite right angles form a long rhombus. We use it to build half of a decagon in one of its corners.

TD2 This part is a bit tedious, but following the circles shown here, we construct the remaining parts of this tessellation. A new thing for us here is the shaded polygon. This is half of a shield from the Topkapi scroll example (see chapter 1)

TD3 Here, we have a variation of the previous tessellation. The decagonal shield was split into a small rhombus and two new polygons. This is not a frequently seen element of decagonal tessellations, but it can be useful while creating patterns in a decagon.

GROUP E. THE TANGENT AND OVERLAPPING DECAGONS METHOD

Tessellations with long hexagons are often difficult to construct. These figures often occur as a part of an arrangement of polygons inside a regular decagon or as an intersection of two decagons. In general, we look for two points that should be the centers of future decagons. We construct two tangent decagons with these centers, and then we produce more tangent decagons.

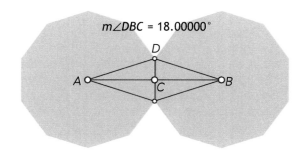

Construction of two tangent decagons

We select two points that we intend to have as centers of two tangent decagons. We construct point C that is a midpoint of segment AB. Using angle 18 degrees, we construct the common edge of the two decagons with both decagons following it.

Here is a very classic example of a tessellation using tangent and overlapping decagons.

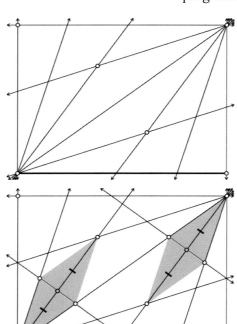

We start with the standard contour and division of the two opposite right angles into 5 equal parts. This is what we did already many times on various occasions. Note – the two intersection points inside the contour. We will use them to make a series of tangent and overlapping decagons.

The two gray rhombi show how we are going to start creating our tangent decagons.

Below left – the first two tangent decagons were created (shaded decagons). They will be overlapping with a two new decagons. Thus remove them, leaving only their outline.

Below right – next, two tangent polygons were created.

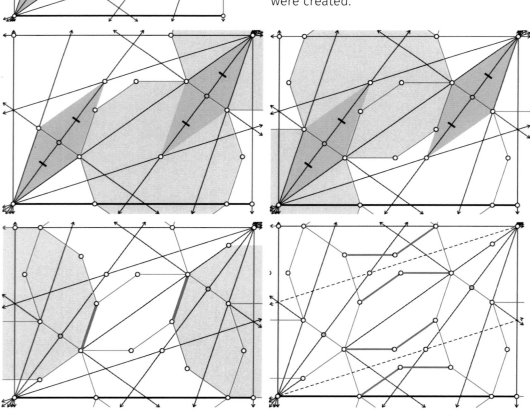

Down left – two new decagons were added. The thick segments show how we created them.

Down right – we finish tessellation near the top and bottom edges of the contour. The thick lines are reflections of the existing lines, and the dashed lines were used as axes of reflection.

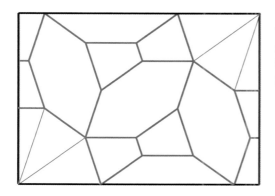

TE1 Tessellation from Vakil Mosque in Shiraz. The pattern created on this tessellation uses very colorful Persian style elements.

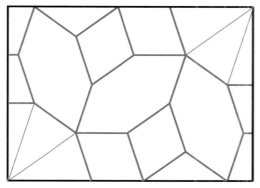

TE2 Variation of the tessellation from Vakil Mosque

This example shows how we could modify the original tessellation to produce a new one. The pattern created using this tessellation will have a completely different look.

IMPORTANT – both tessellations can be treated as starting points for creating more and more complex tessellations.

Our next example was taken from the Akbar Mausoleum in Sikandra, India.
It used a very long contour and six tangent and overlapping decagons.

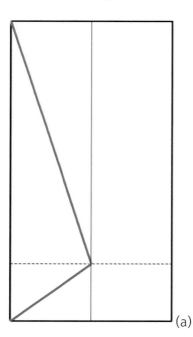
(a)

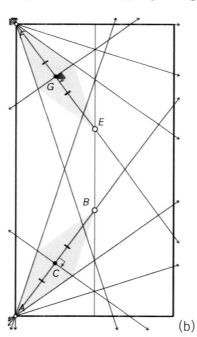
(b)

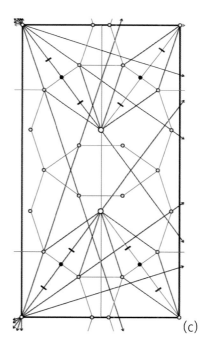
(c)

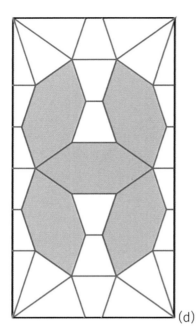 (d) 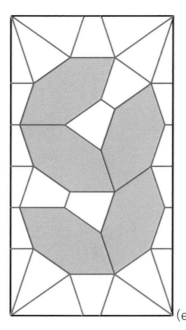 (e) 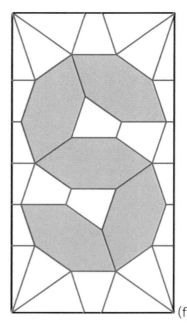 (f)

TE3 Drawing (a) shows how we created contour. Start with the bottom segment, divide it into two equal parts and get the rest as it was shown here. In the drawing (b), we show which points we want to have as centers of the four decagons. This part is completely our own choice. In the drawing (c), we show the decagons created following our choice. The two decagons can be filled with long hexagons in a few completely different ways. Three of them are shown in illustrations (d),(e), and (f).

Patterns created on tangent and overlapping decagons can be very well organized (e.g. (f)), or they may look completely chaotic. There are many mosaics in Iran using this technique, and some of them make an impression of being completely disorganized.

In the drawings above, the tessellation (e) was taken from Akbar mausoleum in Sikandra, India.

Here is another example borrowed from Itimad-ud Daula's tomb in Agra. It shows how chaotic can be such an approach.

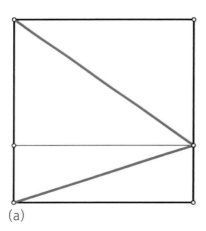 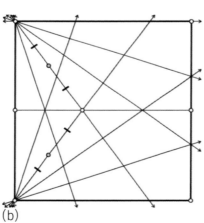 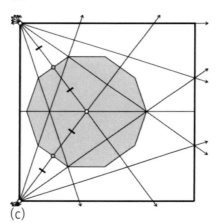
(a) (b) (c)

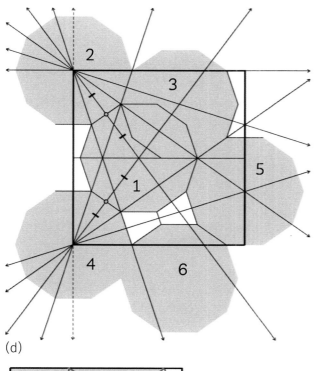

Construction of the tessellation from Itimad-ud Daula's tomb

The creation of the contour is obvious. Then we split it into two equal rectangles (b), and each of the two left corners we divide into 5 equal parts. Drawing (b) shows which points we used to create the first decagons in our tessellation. In drawing (c), we show one of them.

All other decagons in drawing (d) are copies of the first decagon. Numbers show the order of their creations.

This configuration of decagons depends exclusively on our choice. We can try a few other configurations starting from the first decagon. If we succeed, then we will get a framework for a new pattern.

(d)

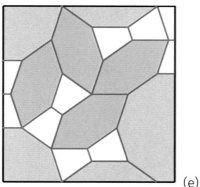

(e)

TE4 The final tessellation

This tessellation looks a bit chaotic, but a pattern created from multiple copies of it will be acceptable.

Note, the half of a decagon on the right edge can be further split into smaller parts. Otherwise, we will have a large rosette on this edge.

The next drawing shows another tessellation created using the tangent and overlapping decagons technique. Its creation is a bit messy, especially the blue part. But it exists.

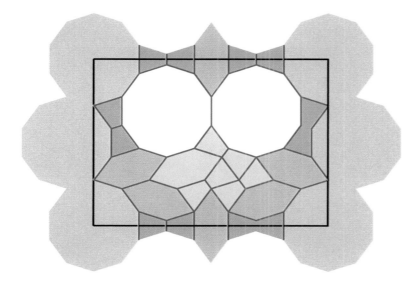

TE5 Another tessellation with long hexagons

The space inside the two white decagons can be filled with a more detailed configuration of long hexagons and tall trapeziums. If we ignore the blue part below the two white decagons, then it is completely symmetric.

More tessellations created with the tangent and overlapping decagons we will see later in a few more examples.

Group F. Tessellations using a single tile

This group of tessellations looks quite strange, but each of them can be used to create a seamless pattern without stars or rosettes. Patterns created with these tessellations are often used as borders to larger and more complex designs. Each of them can be expanded and used in a more complex design. Due to the simplicity of them, we show them here with a pattern filling them.

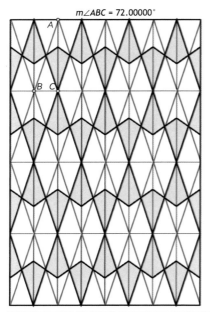

TF1 Tessellation using the long triangle tile

In this tessellation, we have contour C(3/5). This is the vertical version of the contour used in TD2. The rest of this drawing is very obvious. After removing the red segments of the tessellation and thin lines (edges of contours), we will get a seamless pattern that can be used for almost anything from a carving, kundekari design, ceramic tiling to the 3D design with mirrors.

The concept presented here can be applied to any polygon used in decagonal tessellations, excluding a regular decagon and a regular pentagon. Each of them has one symmetry line, and we can split it along the symmetry line and rotate about the center of the right or left side.

Below - tessellation and pattern using flat trapezium as a tile. The pattern is presented in two versions.

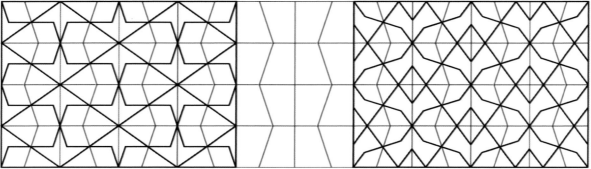

TF2 a tessellation made from a trapezium and sample patterns using it

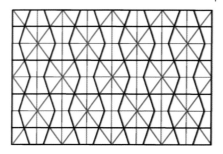

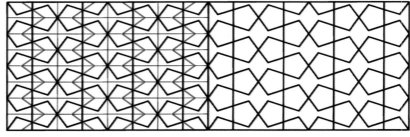

TF3 Tessellation and pattern with rhombus as a single tile.

TF4 Tessellation and pattern with long hexagon as a tile. The right side shows pattern only.

TF5 Here, we have a tessellation created using tall trapezium as a single tile. The drawing shows two versions of this tessellation, left Kukeldash madrasah style and right Persian style.

The top left rectangle shows how this tessellation was created. This is typical contour C(2/5), and the angle shown there is equal to 54 degrees.

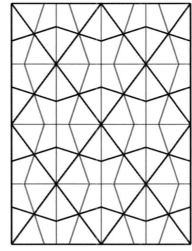

TF6 Tessellation and pattern created using the kite tile

The left-top rectangle shows how the contour and tessellation edges were constructed. The pattern was created by reflections about vertical edges and translations along vertical edges. The vector of translation is equal to the vertical contour edge.

The two horizontal lines show the most convenient horizontal edges of a large design. Otherwise, we will get some small ugly shapes near the top and bottom edge of the pattern.

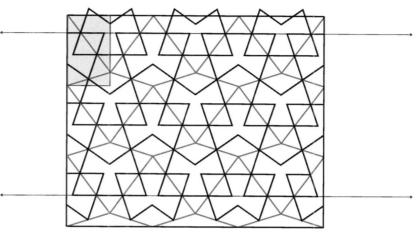

Group G. Tessellation with narrow rhombi

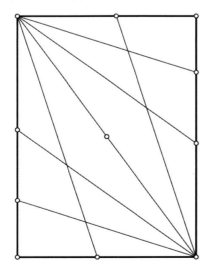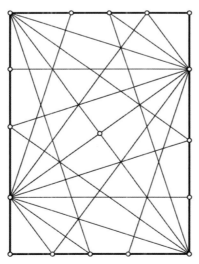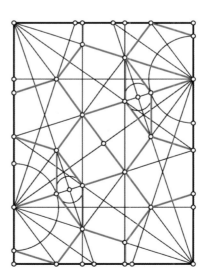

TG1 – Tessellation for the monster pattern Yazd style (project 4.9).

Despite some difficulties in using tessellations with narrow rhombi, they are very valuable in creating interesting designs. They can be used in Yazd style designs. They are also very convenient in designing many decagonal patterns that occur in Cairo mosques.

PROPERTIES OF POLYGONS USED IN DECAGONAL TESSELLATIONS

> **MEMO**
> There exist uncountable quantities of tessellations. It is not possible to show even a small percentage of them. Thus, consider collecting interesting tessellations. This can be a very useful collection.

Each of the polygons that we have seen in this chapter can be constructed from a regular decagon. The figure below[2] shows the relations between each of these figures and a regular decagon.

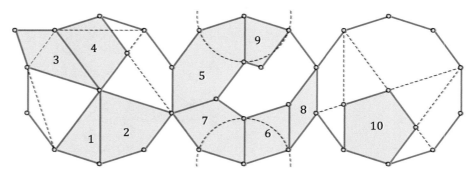

Note many of the polygons included here may occur in two versions, a small version and a large version. For example, the rhombus (nr. 6) may occur in tessellation in a small version (all sides S) and in a large version (all edges L). The polygon with the number 1 may occur in the very large version, sides X and L. Polygon 5, if in the large version, usually can be split into two rhombi (nr. 6), trapezium (nr. 4), both in smaller versions. All depends on the context of a polygon in the tessellation, and each size of a polygon forces a different pattern design. More – some of these polygons may work well in one style of design and make design impossible in another style. In the next chapter, we will talk about these issues.

---

[2] In this figure, we included only some of the most important polygons that we have seen in this chapter. There are a few other decagonal polygons that are not shown here.

6. Kukeldash Madrasah and Persian styles

While walking through mosques, visiting museums, and checking books on Islamic art, we can notice that only two styles mentioned in the previous chapters gained significant popularity. These are the Kukeldash Madrasah style and Persian style. All the remaining styles exist only locally in a few places in the Islamic world. There are many reasons. Some styles are convenient for woodworking, some others for stone or ceramic tiling. Although we got excellent results with the basic tessellation, some styles may not look well when we render them using different tessellations. Some styles simply don't like particular decagonal tiles. Finally, not all styles used on selected tessellations will look charming enough. There is also a matter of personal preferences. We like some shapes, and we dislike some others.

Let us start by looking at the relations between patterns on some tiles and the consequences of the tile sizes used in tessellations. We will start with the Kukeldash Madrasah style and tiles that we used in the basic decagonal tessellation. This will be our base for further investigations. Then we will describe patterns for other decagonal tiles in relation to the pattern on the base tiles. In every case, we have to follow gereh rules – G4, G5, and G6. Especially rule G5 about the continuity of lines passing through the common edge of two tiles will be important. Lines must flow without breaks and bending on the edges of tiles.

Elements of the Kukeldash Madrasah style

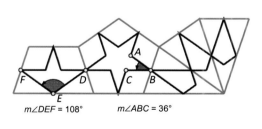

Fig. 6.1. Polygons and the patterns we have used so far.

The pentagon and low trapezium will be our base for further investigations. The trapezium has three edges, S-size, and one L-size. The long triangle has two edges, L-size, and one S-size. A regular decagon can be considered as a union of 10 long triangles. Thus, our regular decagon has all edges S-size.

Later we will deal with regular decagons with L-size edges.

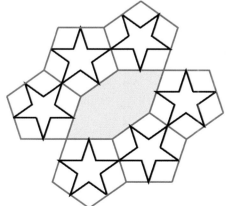

Fig. 6.2. For a start, let us consider a long hexagon with short edges. If we place around it pentagons with the star that we have seen in Kukeldash Madrasah style, we will get the picture shown here. Now, the main question is, what will happen if we extend the stars' edges towards the hexagon's interior? Are we able to create a compatible decoration for the hexagon?

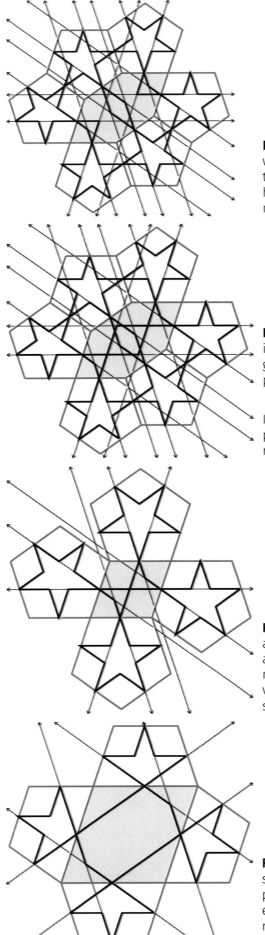

Fig. 6.3. After extending the edges of the stars, we get the network of lines shown here. Now we can try to produce a pattern for the hexagon. Unfortunately, the result that we get here is unsatisfactory, and the big gaps inside the hexagon do not look well.

Fig. 6.4. We could extend the lines of the pattern further like it is shown here. But this time, the pattern inside the hexagon gets unnecessarily crowded, and some shapes between the pattern lines are too small while others are very large.

Important conclusion: any decagonal tessellation using this particular hexagon is not convenient for the Kukeldash Madrasah style pattern design.

Fig. 6.5. A rhombus with short edges is much easier to fill with a pattern in Kukeldash Madrasah style. At least we do not get a gap between the pattern lines. However, while looking at many designs, we will find that this particular shape occurs very rarely. There are very few patterns using a rhombus with short edges in Kukeldash Madrasah style designs.

Fig. 6.6. This time we have a rhombus with L-size edges. Our situation changes completely. The wide angle allows us to produce a large convex hexagon inside the rhombus. However, this shape is too large and does not look nice. For this reason, in most of the designs using this particular rhombus, some additional decorations were added.

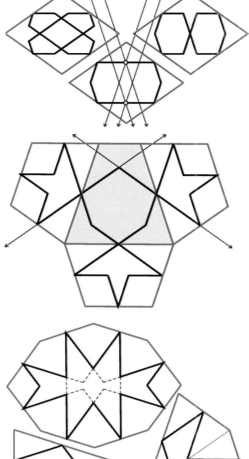

Fig. 6.7. Three different versions of a pattern for the large rhombus. The bottom one shows lines parallel to the sides of this long hexagon. In some designs, these lines were used to add extra decoration for the rhombus and tiles around the rhombus.

Fig. 6.8. The tall trapezium has three edges L-size. Thus, by matching the pattern on the low trapezium, we get this design. The thin lines show directions in which we can still extend the pattern, if we have the right conditions. Otherwise, we can ignore them.

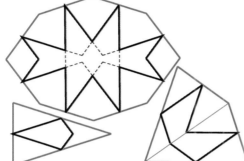

Fig. 6.9. A decagonal shield with S-size edges can be covered with the pattern the way shown here. Usually, the simple version, without the dashed lines, is used.

Below on the left, we have a single long triangle filled with a pattern. This is 1/10 of the pattern in a regular decagon.

The right drawing shows how we can construct a pattern when we have two long triangles side by side in a tessellation. It is a simple variation of the pattern for a long triangle.

Our investigations of the Kukeldash Madrasah style lead us to the conclusion:

> While developing tessellations for Kukeldash Madrasah style patterns, we should avoid using the long hexagon tile. Tessellations with a large rhombus allow us to make some interesting variations of the same design.

In the following projects, we will investigate a few simple patterns rendered in Kukeldash Madrasah style. We will be dealing with some existing patterns. Our goal will be to investigate the pattern structure, construct its contour, guess and construct its tessellation, and then use our knowledge about elements of the Kukeldash Madrasah style and reconstruct the whole pattern. One can go back to the chapter with samples of tessellations and try each of them, if suitable, to construct a pattern on it.

➢ Project 6.1 – Pattern from the Cem Sultan Mausoleum

There are many patterns developed in the Kukeldash Madrasah style. We will use a very simple pattern from the doors to the Cem Sultan mausoleum in Bursa for a start. We have all the tools necessary to reconstruct this pattern and then build something larger based on it.

Fig. 6.10. Simple pattern from the bottom part of the entrance doors to the Cem Sultan mausoleum in Bursa, Turkey.

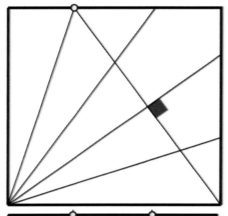

Fig. 6.10a. We start from the very simple, almost square contour. It is the same one that we constructed in the previous chapter. For our design, we will need two copies of it, but it is more convenient to use a single one for the sake of simplicity.

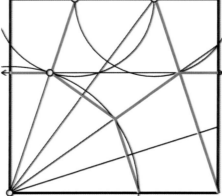

Fig. 6.10b. Here we have the tessellation created on this contour. We made it already in the previous chapter. The points shown here are used to draw the three circles and the supporting lines.

I hope you recognized tessellation TA11 from the previous chapter.

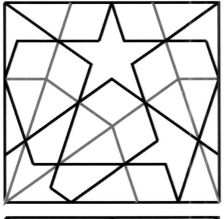

Fig. 6.10c. This image shows the tessellation and the pattern filling it. Some of the tiles, as well as pattern filling it, were cut along their symmetry line by the edge of the contour.

Fig. 6.10d.e. (below) The final template for the pattern from the doors to the Cem Sultan mausoleum. To produce the whole pattern, we will have to turn this template upside down and then add to it its right reflected copy. That is all for this particular pattern.

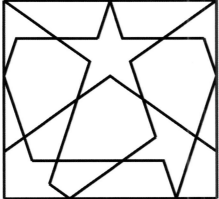
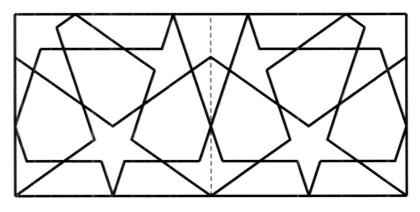

One can also use the same template to produce a larger pattern that is often used for kundekari doors in many places in the Islamic world.

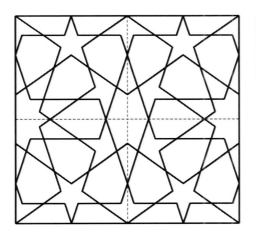
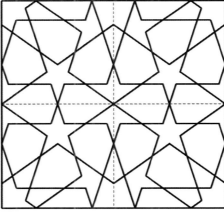

Fig. 6.11 Two slightly different patterns created from the template used for the doors in the Cem Sultan mausoleum

➤ Project 6.2 – Pattern from the Ankara Museum

A very interesting pattern using the template from the previous project we can find in Ankara Ethnographic Museum. Here is its final reconstruction. Can you try to produce this pattern step-by-step?

Fig. 6.12 Reconstruction of the pattern from Ankara Ethnographic Museum. The pattern was created in Kukeldash Madrasah style. Its tessellation uses only decagons, pentagons, trapeziums, and one long triangle. The pattern from each tessellation tile is the same one we described at the beginning of this chapter.

Your tasks will be:
1. Split the pattern into separate instances of the template.
2. Investigate how each instance of the template was produced.
3. Investigate why all these templates fit well together? Hint – check the lengths of the tile edges that are tangent to the top and bottom edge of the contour.
4. Draw the final pattern.

➤ Project 6.3 – Pattern from stone in Amasya

Many mosques in Muslim countries may please us with a nice and interesting pattern. The pattern for this project I found in Amasya in Turkey. The pattern was rendered as a carving in marble and recently was renovated. Thus, we have a very clear design to analyze.

Fig. 6.13. A decagonal pattern in the form of a long band in the Beyazid Paşa Mosque in Amasya.

Our task will be to analyze this pattern and use it in our own design.

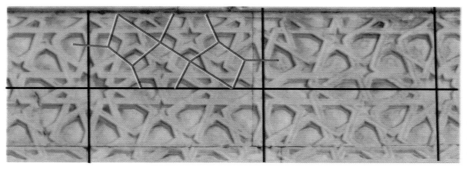

Fig. 6.13a. In this drawing, we have the same pattern divided into rectangles by its symmetry lines (black). One of the rectangles was split into a tessellation. Note – here, we have the same tiles that we had in two previous projects. Only the contour and tessellation are different.

This example shows us that many patterns that look very complicated are, in fact, simple modifications of the basic Kukeldash Madrasah pattern that we developed at the beginning of this book.

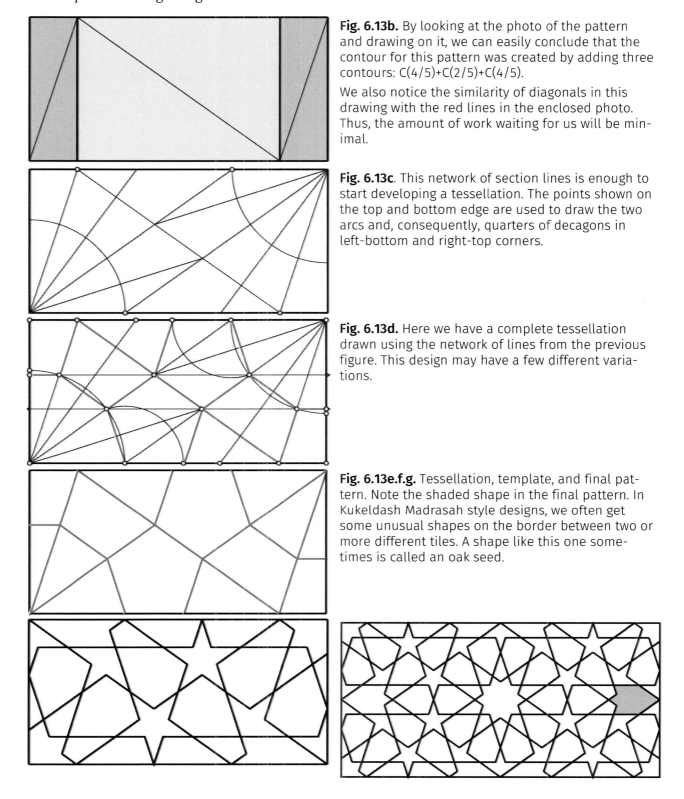

Fig. 6.13b. By looking at the photo of the pattern and drawing on it, we can easily conclude that the contour for this pattern was created by adding three contours: C(4/5)+C(2/5)+C(4/5).

We also notice the similarity of diagonals in this drawing with the red lines in the enclosed photo. Thus, the amount of work waiting for us will be minimal.

Fig. 6.13c. This network of section lines is enough to start developing a tessellation. The points shown on the top and bottom edge are used to draw the two arcs and, consequently, quarters of decagons in left-bottom and right-top corners.

Fig. 6.13d. Here we have a complete tessellation drawn using the network of lines from the previous figure. This design may have a few different variations.

Fig. 6.13e.f.g. Tessellation, template, and final pattern. Note the shaded shape in the final pattern. In Kukeldash Madrasah style designs, we often get some unusual shapes on the border between two or more different tiles. A shape like this one sometimes is called an oak seed.

Before moving to the other design styles, let us look at another example of a Kukeldash Madrasah design.

➤ Project 6.4 – A pattern from the Beyazid Mosque Amasya

Fig. 6.14a.b. The pattern from the old doors in Amasya shows a very strange shape located on ¼ and ¾ of its height. In the right image, we have a hand drawing of the tessellation of this pattern. Here we can clearly see that the strange shape is a simple pattern filling the shield we discussed before.

Our goal will be to create a very specific contour used for this pattern and its tessellation.

Note – this is precisely the same pattern from the Topkapi scroll that we have seen in chapter 2.

From the picture, it is not obvious how the contour for this pattern was created. It is very specific, and it is a combination of two different contours.

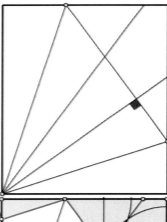

Fig. 6.14c. One of the contours used for this pattern. We will need two such contours placed, one above the other and a thin C(1/5) contour between them. The middle contour will allow us to do the translation between two different shields.

Fig. 6.14d. Tessellation for this contour can be created the same way as we did in the first project in this book.

Note the two shaded polygons. On the right one this it is a quarter of the shield cut along its symmetry lines. The left one is a narrow fragment of the same shield that we see on the whole tessellation's left side.

To connect these two shields properly, we will need an extra contour C(1/5).

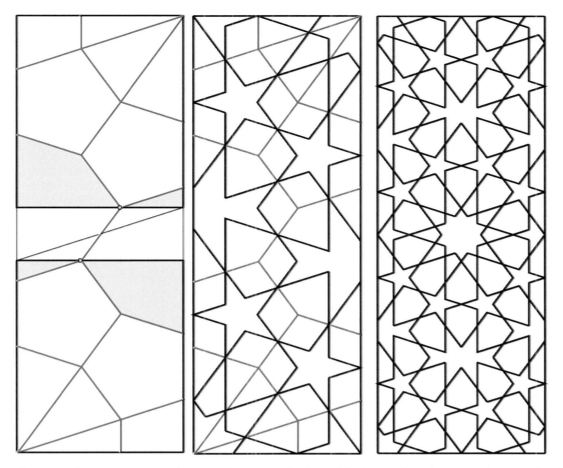

Fig. 6.14e.f.g. Here we see the complete tessellation with the enclosed connection part in the middle; tessellation and the final pattern from Amasya. The pattern for the shield can be more elaborated, but then some areas of the pattern will be unnecessarily crowded.

➢ PROJECT 6.5 – PATTERN FROM DAMASCUS

Fig. 6.15. A decagonal pattern on the side of the minbar in the Great Mosque in Damascus

The Great Mosque in Damascus was decorated with numerous patterns in various styles. We can find there patterns using octagonal, dodecagonal, and decagonal symmetries. The pattern on the minbar is a very simple decagonal pattern using the Kukeldash Madrasah style. The very interesting feature of this pattern is the use of a large rhombus in tessellation for this pattern and how the large empty space inside the hexagon was filled with extra decoration.

In this project, we will investigate this pattern and reconstruct it step-by-step.

Important – this project is much harder than all previous projects. You can skip it now and come back to it later.

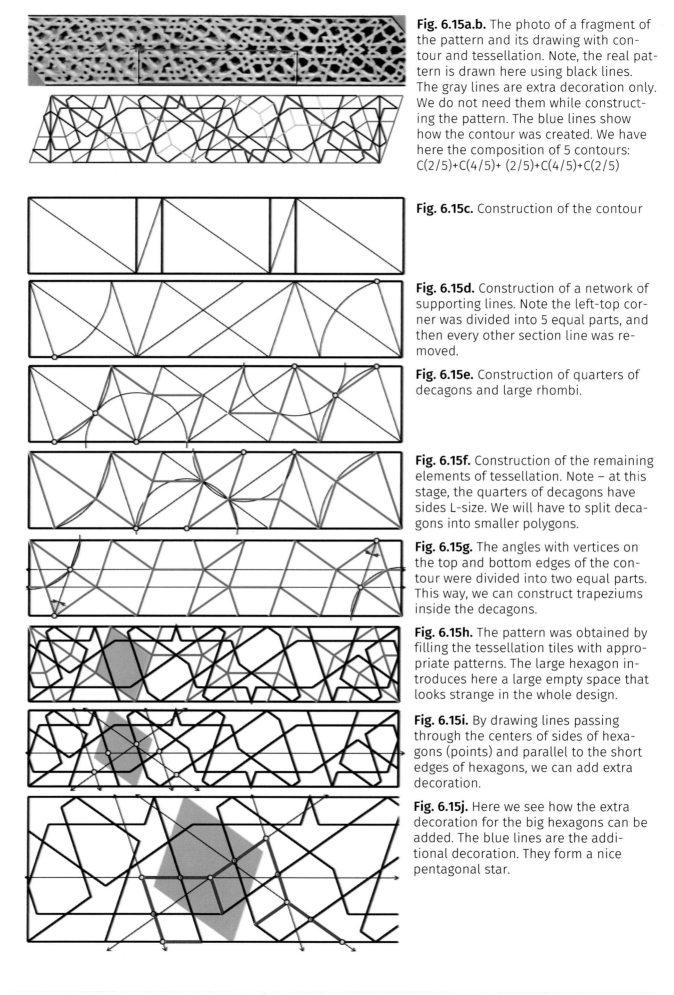

Fig. 6.15a.b. The photo of a fragment of the pattern and its drawing with contour and tessellation. Note, the real pattern is drawn here using black lines. The gray lines are extra decoration only. We do not need them while constructing the pattern. The blue lines show how the contour was created. We have here the composition of 5 contours: C(2/5)+C(4/5)+ (2/5)+C(4/5)+C(2/5)

Fig. 6.15c. Construction of the contour

Fig. 6.15d. Construction of a network of supporting lines. Note the left-top corner was divided into 5 equal parts, and then every other section line was removed.

Fig. 6.15e. Construction of quarters of decagons and large rhombi.

Fig. 6.15f. Construction of the remaining elements of tessellation. Note – at this stage, the quarters of decagons have sides L-size. We will have to split decagons into smaller polygons.

Fig. 6.15g. The angles with vertices on the top and bottom edges of the contour were divided into two equal parts. This way, we can construct trapeziums inside the decagons.

Fig. 6.15h. The pattern was obtained by filling the tessellation tiles with appropriate patterns. The large hexagon introduces here a large empty space that looks strange in the whole design.

Fig. 6.15i. By drawing lines passing through the centers of sides of hexagons (points) and parallel to the short edges of hexagons, we can add extra decoration.

Fig. 6.15j. Here we see how the extra decoration for the big hexagons can be added. The blue lines are the additional decoration. They form a nice pentagonal star.

90 | Practical geometric pattern design – decagonal patterns in Persian traditional art

Fig. 6.15k. Final pattern from the Great Mosque in Damascus.

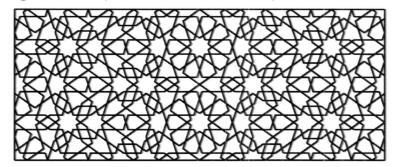

Fig. 6.15l. Here we have another version of the pattern from the Great Mosque in Damascus. Can you find out what is different and how this pattern was created?

Fig. 6.15m.n. Decagonal rosette using a triangular fragment of the pattern from Damascus. Below is shown the triangular template for the rosette.

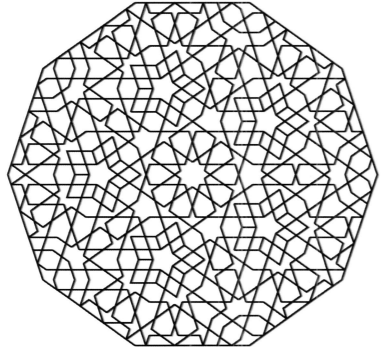

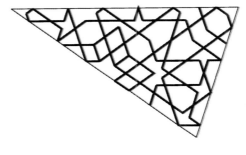

Patterns in the Kukeldash Madrasah style form a huge and very diverse group. There are a few techniques allowing us to create some incredibly complex Kukeldash Madrasah style patterns. We will come back to them in another book. Due to their complexity, they go much beyond a book for complete beginners.

Elements of the Persian style

As we have seen in the previous section, some of our decagonal polygons are not suitable for designing Kukeldash Madrasah style patterns. The same situation we have for the Persian style. Let us look at what may go wrong this time. We know how the pattern is constructed for a pentagon, decagon, and trapezium. Now we will have to elaborate the pattern for each of the remaining tiles.

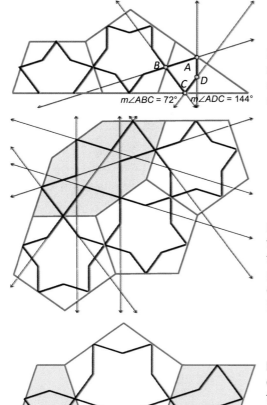

Fig. 6.16. We know well the pattern for the pentagon, long triangle, and trapezium. In fact, in our Persian style example, we did mention that the pattern for a trapezium can be more elaborate. We will have to look into this issue later.

Fig. 6.17. Construction of pattern for the long hexagon. Here, the pattern lines are natural extensions of lines of the pattern from stars around the hexagon. Note all the edges of all polygons here are S-size. The long hexagon usually does not occur in decagonal tessellations in L-size. The reason is that it can be easily split into other smaller tiles.

Fig. 6.18. The two other polygons with S-size edges can be covered with a pattern without any problem. Here we see how the tall trapezium and rhombus can be filled with patterns. We need to follow the extensions of lines of the pattern from a pentagon attached to them.

Generally, in the classic Persian style, we deal with tessellations using polygons with all edges S-size or shorter. The only exception is a low trapezium where three edges are S-size and one L-size. Decagons are considered as a whole, and the long triangle does not occur outside of decagons as a separate tile or as a tween with another long triangle.

In the series of following projects, we will investigate a few simple patterns rendered in Persian style. This time we will be dealing with some tessellations introduced in the previous chapter. Our goal will be to use our knowledge about elements of the Persian style and construct the whole pattern. Any Iranian book dealing with geometric patterns contains many examples of classic Persian patterns (see, for example, Heli, 1986).

➤ PROJECT 6.6 – PATTERN FROM BEYAZID II MOSQUE IN ISTANBUL

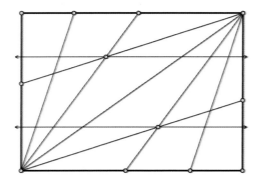

Fig. 6.19
Let us start with a tessellation that was introduced in the previous chapter as TA5. It uses contour C(2/5), and network lines show on the drawing.

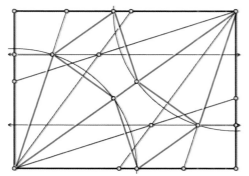

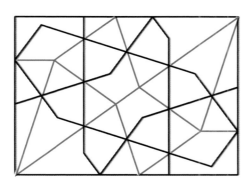

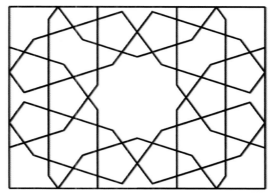

Fig. 6.19b.c.d. Construction of the tessellation, below we see the template for the pattern and a larger piece of the pattern using four copies of the template.

We already created this tessellation as TA5.

➤ Project 6.7 – Pattern from Muradiye Mosque in Bursa

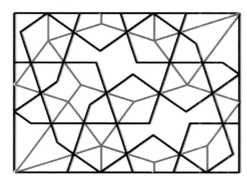

Fig. 6.20.a.b. Tessellation T10 was one of the complex tessellations in the previous chapter. For the exact construction of it, check the drawings that we did before.

Below I show a complete template of the pattern and a simple pattern created using sixteen copies of the template. It looks quite complicated, although its template was reasonably simple.

Note – the original pattern in Bursa uses slightly different tessellation.

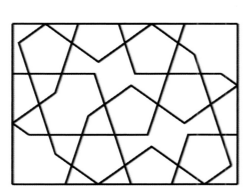

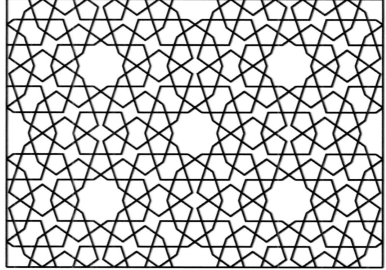

➤ Project 6.8 – Tessellation with four hexagons

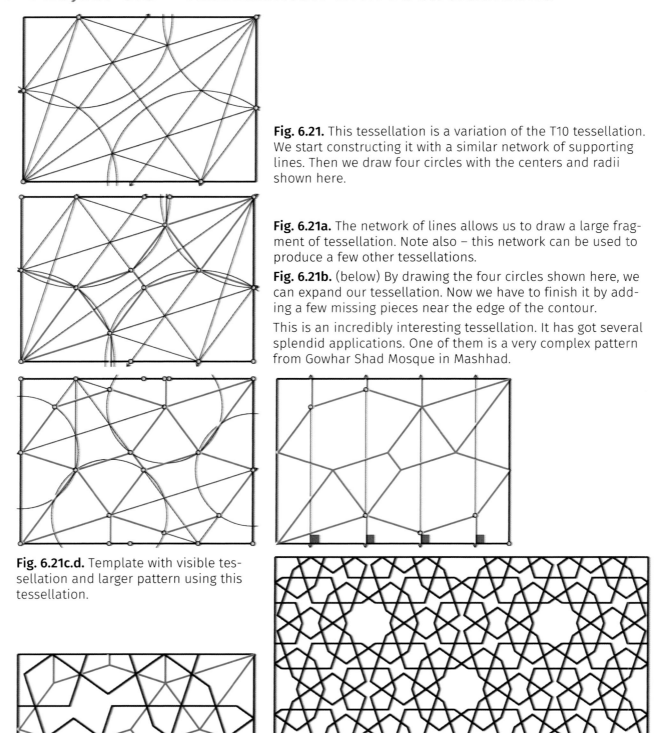

Fig. 6.21. This tessellation is a variation of the T10 tessellation. We start constructing it with a similar network of supporting lines. Then we draw four circles with the centers and radii shown here.

Fig. 6.21a. The network of lines allows us to draw a large fragment of tessellation. Note also – this network can be used to produce a few other tessellations.

Fig. 6.21b. (below) By drawing the four circles shown here, we can expand our tessellation. Now we have to finish it by adding a few missing pieces near the edge of the contour.

This is an incredibly interesting tessellation. It has got several splendid applications. One of them is a very complex pattern from Gowhar Shad Mosque in Mashhad.

Fig. 6.21c.d. Template with visible tessellation and larger pattern using this tessellation.

Tessellations with long hexagons, regular decagons, rhombi, and tall trapeziums were frequently used in Persian style patterns. In the next project, we will construct a very complex tessellation using three tiles only – regular decagon, long hexagon, and tall trapezium. The last polygon I often call 'filler' as it often fills gaps between other polygons. For this project, we will use a method that I invented some time ago while researching Persian patterns. For the method, I use the name 'method with tangent and overlapping decagons.' The method is quite powerful and flexible and can be expanded to a few other types of geometric patterns. The method of tangent and overlapping decagons was presented in the previous chapter while discussing tessellations in group E.

➤ PROJECT 6.9 DOORS FROM THE DAVID COLLECTION

The pattern for this project can be seen in the David Collection Museum in Copenhagen in Denmark.

Fig. 6.22 Fragment of wooden doors from the C. L. David Foundation and Collection in Copenhagen.

The doors shown here are believed to have been created in the XVII century in Iran.

By analyzing the pattern, we can easily distinguish the shapes that we have seen in previous projects. The most often used are the patterns filling the long hexagon and tall trapezium (the filler). In the corners and center, we see decagonal stars typical for all Persian style patterns that we have seen in this chapter.

Our goal, as in other projects, will be to

Construct the contour

Construct tessellation and

Produce the pattern using shapes that we developed already.

We will also need to investigate how one can fill a regular decagon with the tessellation using the three shapes already mentioned: long hexagon and filler.

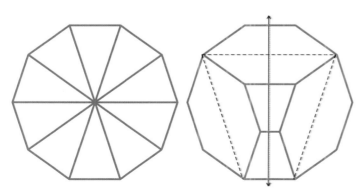

Fig. 6.22a. In this figure, we have presented two ways of filling a regular decagon with a tessellation of long triangles (part of a decagon) and a long hexagon and filler.

The first one has 10 mirror lines and usually is located in the corner of a contour. Sometimes in other places. The second tessellation of a decagon has one symmetry line only. Thus, if we have to cut it by the contour edge, the edge should be incident with this symmetry line or with a symmetry line of one of the long hexagons (dashed lines).

Now, with this knowledge, we can start developing our project.

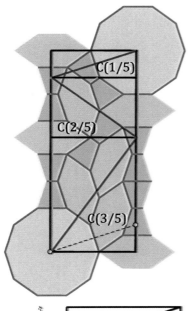

Fig. 6.22b.
This image shows the top-right quarter of the pattern. We can notice that the contour is built out of three stacked contours C(3/5), C(2/5), and C(1/5).

I obtained the shown-here tessellation by drawing lines on the photograph. Note, by drawing it, we get some idea about how our construction should look, but drawing or assembling puzzle tiles does not mean that we know how to construct such patterns. Unfortunately, many people think assembling puzzles using some predefined tiles is the default or original method for constructing geometric patterns.

We will have to produce an algorithmic method for developing such and similar tessellations. Even in this contour, we can produce a few different tessellations using decagons, long hexagons, and fillers.

The key points for our observations are the two decagons at the bottom, with their centers marked by small circles. This is where we have to start. The two points define the size of all the decagons used in this project. All of them will have the same size.

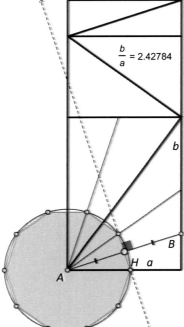

Fig. 6.22c. This is the beginning of the construction. We construct the contour the same way as we did many times before. Then we divide the left-bottom corner into 5 equal parts. The bottom section line AB we divide into two equal parts (the large point is the midpoint). By drawing a line (dashed) passing through the midpoint and perpendicular to this section line, we can produce the left decagon.

Now is the essential task. Take some tracing paper and copy this decagon with its corners and the center. In the next few steps, we will use it many times. Thus, your tracing copy of the decagon should be perfectly done.

By the way – all examples like this one should be done as large as possible. Thus, if you want to draw this quarter of the pattern on A4 paper, you should calculate its size precisely. For example, if the long edge of the contour will be 25 cm, then the short edge should be $25/2.42784=10.29≈10.3$

In such a case, the radius of the decagon will be around 5.7 cm. This is enough for such a simple pattern like this one. But it may not be enough for more complex patterns.

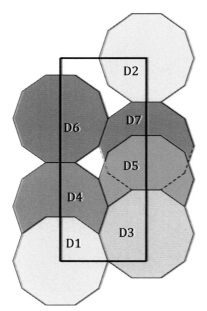

Fig. 6.22d. Now, we make 7 copies of the decagon in the order shown here. Decagon D1 is the original. Decagon D2 is its copy, and it should be placed with high precision. If we forget about it, then we can draw it at the end of the drawing process.

Decagon D3 is tangent to D1. D4 is tangent to D3, D5 and D6 are tangent to D4, D7 is tangent to D6 and D3. D2 is tangent to D7.

Important – each new decagon goes below all previous ones.

Thus D4, D5, and D7 are partially hidden by D1 and D3.

In the next step, we will have to fill the visible part of each decagon with patterns. Decagons D1 and D2 should be filled with the star pattern – here, we do not have much choice if we want to follow the original pattern. The other decagons we will with fill according to the figure 6.22a. We do not have many options here. These decagons are cut by the contour lines. Thus, the edge of the contour should match appropriate symmetry lines. The only place where we have some choice is the D6 decagon.

IMPORTANT: If we draw decagons in a different order, we will simply get another pattern.

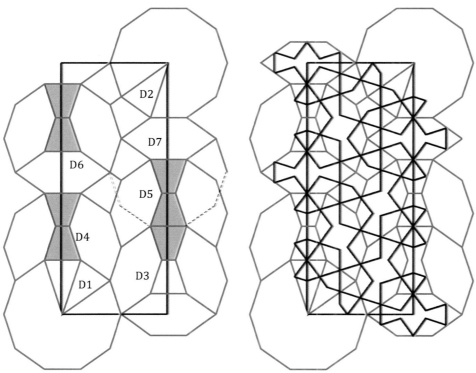

Fig. 6.22e.f.g.
Left: Tessellation for the pattern for the khatamkari doors from David Collection.
Right: Tessellation and the pattern
Next page: final template (left) and the final pattern (right).

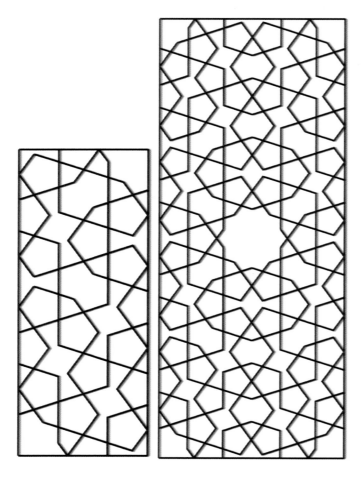

Now we can experiment with the pattern from the doors from David Collection. If we choose a different decagons creation order, we will get another pattern in exactly the same style.

A contour with C(3/5) and C(2/5) will give us a similar pattern but slightly shorter.

➤ Project 6.10 An experimental Persian design

In the previous chapter, we discussed some specific contours and tessellations. Let us take one of them and see how we can use it for designing our own pattern. For this project, I selected tessellation TC8. It is a union of three tessellations.

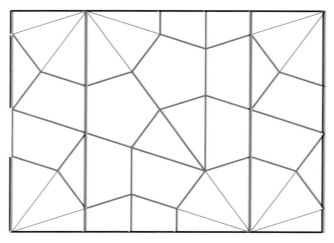

Tessellation TC8

We know all shapes used in this tessellation. It can be used to create a Persian style pattern, but we will have to create a new pattern for the two long triangles in the center of this drawing.

Let us proceed with all shapes that we already know and their patterns.

The two long triangles problem

This drawing shows what will happen if we fill the two encircled triangles using a typical for them pattern. The large empty gap that looks like a big bird or human dress does not occur in any Persian style pattern.

But there is an easy way to fix this problem.

A fill for two tangent triangles

The construction below shows how we could create a pattern filling two tangent long triangles. This is often seen solution in decagonal designs all over Iran.

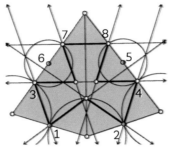

A large pattern for this design

The drawing shows one of many patterns that we can create using the template shown above left. We could have more or fewer copies of this template and obtain a few interesting designs. There is one problem with these designs. The bog stars scattered all over the pattern leave large empty gaps. We often may not want to have them. A good way out is to create a partial template with decagons left empty; create a large design using this template and then fill the empty decagons in any way we want.

6. Kukeldash Madrasah and Persian styles | 99

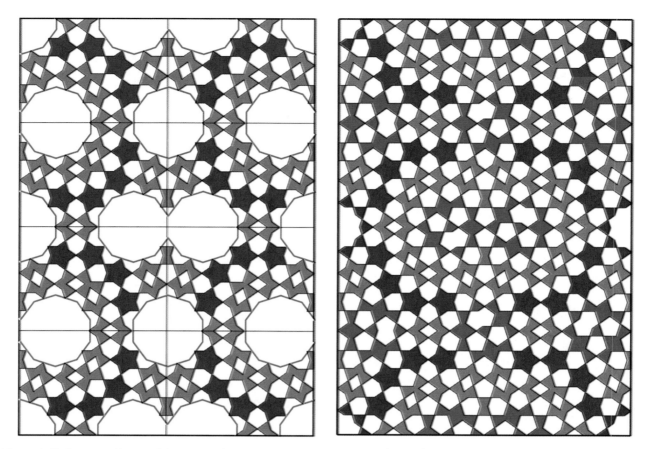

Above left, large pattern with empty decagons. Now we can use these decagons to fill them with a pattern. Look back into project 6.10 and the ways for filling decagons. The final design may look very well organized or completely chaotic (image above right).

Many interesting examples of decagonal patterns from Persia and Central Asia are worth discussing and enjoying their creation. We can return to them later – in a new book or a series of papers. This last example was a glimpse into the next book on more advanced Persian decagonal geometric patterns. The book entitled "Practical Geometric Pattern Design – lessons from Topkapi scroll" will be available later this year. We will discuss complex patterns and methods to increase pattern complexity. Many patterns discussed there will come from the Topkapi scroll and Mirza Akbar Khan scrolls. But this is not the end of the story of decagonal patterns. There are still hundreds of interesting examples, ideas, and methods worth investigating. So, see you again soon.

Bibliography

The literature for geometric patterns in Islamic art is very wide-ranging. Numerous books and papers discuss various features of these patterns and various aspects of the design of these patterns. This bibliography gives a very limited selection of books and papers used while researching and writing this book. For each position, I give a very short remark about the content and the importance of it.

D'Avennes, P., (2007), Islamic Art in Cairo from the Seventh to the Eighteen Centuries, with an introduction by George T. Scanlon, A Zeitouna Book, The American University in Cairo Press, Cairo – New York.

> This book is a reprint of the monumental book L'art arabe d'apres les monuments du Kaire, published serially by Prisse d'Avennes between 1867 and 1879. Readers will find here numerous drawings of Islamic architecture that still existed in Cairo during Avennes times. Most of the drawings are very accurate, and the majority of them are rendered in color.

Bourgoin J., (1973), Arabic geometrical patterns & design, Dover Publications, Inc., New York.

> This is a reprint of the book 'Les Elements de l'art arabe: le trait des entrelaces', originally published by the Librarie de Firmin-Didot et Cie, Paris, in 1879. This book is an excellent source of geometric patterns in Islamic art from Egypt and neighboring countries. Most of them are very accurate, and the range is quite wide. There are no descriptions of how these patterns were created, and all illustrations are black-and-white. In the original book, the plates were printed in color. Many of the patterns from this book do no longer exist.

Bulut M., (2020), Selçuklu Çizgileri-Anadolu Selçuklu Geometrik Kompozisyonları

> M. Bulut is one of the best specialists in Seljuk architecture. The book contains the largest collection of drawings of Seljuk designs from Anatolia. The drawings are very accurate.

Grünbaum B., Shepard G.C. (1989), Tilings and Patterns an Introduction, W.H. Freeman and Company, New York.

> This is the major monograph on tilings and patterns based on tilings. This book is an essential volume for Geometric patterns in Islamic art designer. Many of the types of tilings presented in this book can be used or were already used for creating Islamic art.

Heli S. A., (1986), Gereh & Arc. Kashan, Iran

> Full name of the author Sayed Akbar Helli. The author concentrates on two major topics: constructing window arcs in Persian architecture and geometric patterns.

Lee T., (1975). Islamic Star Patterns – Notes, unpublished manuscript available online as PDF files from http://www.tilingsearch.org/tony/

> This is an excellent source of ideas related to Islamic art's geometric patterns, their specific features, and some construction details. I consider it as the most reliable source on the geometry of patterns with a rigorous mathematical approach.

Majewski M., (2020). Practical Geometric Pattern Design – Geometric Patterns from Islamic Arts

> This book is an expanded version of a textbook for one of the author's courses at the Istanbul Design Center. It introduces various types and methods of pattern design. It can be treated as a first source of information for a beginner in geometric pattern design.

Rempel, L.I., (1961), Architectural Ornament of Uzbekistan, National Publisher Artistic Literature UzSSU (in Russian).

The book was published in Russian. The original reference for this book is: Ремпель, Л.И., (1961), Архитектурный Орнамент Узбекистана, Государственное Издателство Художественной Литературы УзССР, Ташкент. The author is another legendary person in Russian archeology and the history of the art of Central Asia. As an exile and Jew, he was prohibited from having full-time employment in the whole SU. Thus his works should be considered as an expression of his particular passion.

Although the book was published in Uzbekistan in 1961, it is no longer available on the market due to the political changes in the former Soviet Union. There are very few copies in major libraries in Russia as well as in the West. This monograph gives a very interesting historical overview of Islamic ornaments in Central Asia. In this book, we will find a few geometric ornaments that are specific to Central Asia designing centers.

Wade D., (1976), Pattern in Islamic Art, The Overlook Press Woodstock, New York.

Wade's book is one of the first books explaining geometry lying behind Islamic art's geometric patterns. The author shows the construction of many patterns that can also be found in Bourgoin's book 'Les Elements de l'art arabe: le trait des entrelaces' (1879). An interesting feature of this book is a large collection of interlaced patterns.

Wade D., (1976a), Pattern in Islamic Art (http://patterninislamicart.com/)

A website by David Wade contains over 4000 photographs of Islamic patterns all over the World. This is the most extensive collection of photographs of Islamic patterns ever created. They can be very useful while analyzing drawings from Bourgoin (1973).

Wichmann B., Tiling Database (http://www.tilingsearch.org/).

Brian Wichmann's tiling database is a unique database of geometric patterns created with segments. In this database, no patterns are using circular arcs or circles. The database is connected, where it is possible, with David Wade's pattern in the Islamic art archive. All patterns were created with the computer software created in cooperation with Anthony Lee.

Made in the USA
Middletown, DE
28 September 2023

39680121R00060